SCARCITY BRAIN

ALSO BY MICHAEL EASTER

The Comfort Crisis

SCARCITY BRAIN

Fix Your Craving Mindset and
Rewire Your Habits to Thrive with Enough

MICHAEL EASTER

RODALE

NEW YORK

Published in the United States by Rodale Books, an imprint of Random House,
a division of Penguin Random House LLC, New York.

RodaleBooks.com | RandomHouseBooks.com

RODALE and the Plant colophon are registered trademarks of
Penguin Random House LLC.

Library of Congress Cataloging-in-Publication Data
CIP data is available upon request.

ISBN 978-0-593-23662-8

Ebook ISBN 978-0-593-23663-5

Printed in the United States of America

Book design by Andrea Lau
Art throughout by tele52/Shutterstock.com
Jacket design by Pete Garceau
Jacket art by Galyna_p/iStock/Getty Images Plus

10 9 8 7 6 5 4 3 2 1

First Edition

To Leah, obviously

CONTENTS

INTRODUCTION

Our Scarcity Brain

Qutaiba Erbeed, my fixer in Iraq, is the most full-of-crap person I've ever met. That's how he'd fast-talked us into a fortified police compound on the outskirts of Baghdad.

We were sitting on a hardwood bench in a makeshift waiting room. Photos of terrorists and drug kingpins filled the wall behind us. Each picture showed a man standing in handcuffs with confiscated weapons and chemical compounds all splayed out in front of him. Big bags of pills, bricks of powder, AK-47s, makeshift bombs, rocket-propelled grenade launchers. Captions in Arabic listed the person, the place, the haul.

A closed-circuit TV hanging in a corner displayed a live feed of the holding cells. One heavily guarded cell contained eight of the region's most wanted and dangerous men.

We were waiting to speak with Mohammed Abdullah, Baghdad's head of drug enforcement. Erbeed had lured me to Iraq with a detailed "itinerary." It said he'd arranged all sorts of important meetings, one of which was to ride along with Abdullah's A-team as they raided drug and terror cells.

But after four days in Iraq, nothing had happened. When I'd arrived and paid him, Erbeed admitted, "The itinerary was . . . um . . . *proposed*. Yes, it was a *proposal*."

But now, it appeared, Erbeed had perhaps talked Baghdad's finest into letting this ride along happen. "They say it's okay, but we must wear flak jackets," Erbeed said, pleased with himself. "We now wait for the final response."

As we sat, Iraqi narco detectives—plain clothes, thick mustaches, with pistols in the waistbands of their jeans—emerged from offices like coyotes, all trying to sniff out why this gangly American was sitting in their waiting room. They circled me but didn't engage. Instead, they all chatted, chain-smoked, and side-eyed me.

Eventually one stepped from an office and approached us. He began talking. "Ride alongs?" he said. "Who told you this can happen? No. This cannot happen. Too dangerous."

"How dangerous?" I asked.

"I was shot three times last week," the officer said. Erbeed and I tried to appear unruffled.

"The dealers are becoming more violent," the officer continued. "Many are transporting and selling quantities of drugs large enough that the penalty is the death sentence. So they will fight to escape."

Erbeed and I collected ourselves, huddled, and considered this. We then explained that we accepted the risk and would stay well in the background.

The officer looked me directly in the eye and held it as he methodically pointed to three spots on his chest. "I'd be dead if I weren't wearing a vest last week," he said.

Then he shrugged. "But, okay, I will ask."

He tiptoed to Abdullah's office, knocking lightly on the door and bowing his head as he entered.

Baghdad is generally considered a good place for solo journalists to get kidnapped and sold to ISIS, whatever they're there for. I was there for the drugs.

I was investigating the dramatic rise of a new, methamphetamine-like street drug called Captagon. It's hardly known in the United States, but it's wreaking havoc in the Middle East and spreading. How I ended up in Iraq, however, takes some explaining.

The short answer: it was the pandemic, and I wasn't thinking rationally. But there's a longer answer.

As a science journalist and professor, I'm interested in understanding human behavior. Everyone likes to focus on developing good new habits. But I want to know how we can resolve the behaviors that hurt us most. Because here's the thing: it doesn't matter how much gas we give good new habits; if we don't resolve our bad ones, we still have our foot on the brake.

And I'd begun noticing a unique signature of the behaviors that hurt us most. We can quickly repeat them. The worst habits are things we can do over and over and over in rapid succession—eventually to our detriment. These behaviors are often fun and rewarding in the short term but backfire in the long run.

We all do stuff like this to some degree. And even if we realize that these behaviors have turned counterproductive, we find it hard to stop.

Everyone knows any behavior is fine in moderation. But why do we suck so bad at moderating? Why do we keep eating when we're full? Why do we keep shopping when we own too much? Why do we keep drinking when we're already tipsy? Why do we scroll social media when it makes us miserable? Why do we binge-watch another episode even when we realize a more meaningful life beyond the screen is passing us by? Why do we get stuck? Stuck doing the same thing we regret over and over and over.

I learned that these behaviors are usually reactions to feelings of "scarcity." And all it takes is a small "scarcity cue" to incite them.

A scarcity cue is a piece of information that fires on what researchers call our scarcity mindset. It leads us to believe we don't have enough. We then instinctually fixate on attaining or doing that one thing we think will solve our problem and make us feel whole.

Scarcity cues are like air: all around us and inside us. They can hit us through advertising, social media, news, chats with co-workers, walks in the neighborhood, and so much more. They can be direct and all encompassing, like a sagging economy or global pandemic. Or they can be subtle and slight, like our neighbor buying a shiny new car.

Our reaction to scarcity isn't anything new. It's an ancient behavior system that evolved naturally in the human mind to help our ancestors survive.

Scientists detailed our scarcity mindset and reaction to scarcity cues as early as 1795. And the topic is now an intense area of research for psychologists, anthropologists, neuroscientists, sociologists, economists, and biologists.

Today it's well accepted that for most of human history, obeying the next scarcity cue and constantly craving and consuming more kept us alive. We evolved in harsh environments that had one thing in common: they were worlds of less, of scarcity.

Things critical to our survival like food, information, influence, possessions, time on earth, what we could do to feel good—and much more—were scarce, hard to find, and short-lived. The people who survived and passed on their genes chased more. They defaulted to overeating, amassing stuff and information, seeking influence over others and their environments, and pursuing pleasure and survival drives to excess.

Obeying these evolutionary cravings kept us alive and still makes sense for all species. Except one.

As humans figured out how to make things faster and cheaper during the Industrial Revolution, our environments of scarcity rapidly shifted to those of plenty. By the 1970s, the benefits of this revolution had spread to most people in developed countries. They've been rippling out across the globe ever since.

We now have an abundance—some might say an overload—of the things we've evolved to crave. Things like food (especially the salty, fatty, sugary variety), possessions (homes filled with online purchases), information (the internet), mood adjusters (drugs and entertainment), and influence (social media).

Yet we're still programmed to think and act as if we don't have enough. As if we're still in those ancient times of scarcity. That three-pound bundle of nerves in our skull is always scanning the background, picking up and prioritizing scarcity cues and pushing us to consume more.

We're still compelled to eat more food than our bodies need. To impulsively search for more information. To buy more unnecessary stuff. To jockey for more influence over others. To do what we can to get another fleeting hit of pleasure. To fixate on getting what we don't have rather than using and enjoying what we do have. We have a scarcity brain.

The science shows that our scarcity brain doesn't always make sense in our modern world of abundance. It now often works against us, and outside forces are exploiting it to influence our decisions. It's at the root of the counterproductive behaviors we can't seem to shake. The habits that put a hard brake on improving our physical and mental health, happiness, and ability to reach our full potential. Aren't addiction, obesity, anxiety, chronic diseases, debt, environmental destruction, political dispute, war, and more all driven by our craving for . . . more?

Humanity has experienced big scarcity cues before. But the Covid-19 pandemic occurred at a strange moment. A time when technology has accelerated to deliver abundant access to everything we're built to crave, while also giving corporations unprecedented insight into exactly how they can leverage our scarcity brain to bend our behavior. Especially those behaviors we can repeat over and over and over in rapid succession— eventually to our detriment. It's as if there's some larger behavior pattern at play . . . almost like a scarcity loop. I even started calling this pattern I noticed a "scarcity loop." And it seemed to be the serial killer of moderation.

We may be out of the pandemic, but the wave of craving and consumption it caused hasn't subsided. We've always been moving toward more. And much smaller scarcity cues have always subtly steered our everyday lives. They've pushed us into that scarcity loop behavior pattern of quick repeat consumption in even the best of times.

And that's why I was in this police compound in Baghdad. I suspected that the rise of that new drug Captagon in this treacherous city held implications for the rest of us. It could help us understand what happens when our scarcity brain meets a sudden abundance of a substance that can push us into a scarcity loop—satisfying us in the short term but hurting us in the long run. And from there, I could begin to unpack what we can do about all sorts of counterproductive behaviors.

Iraq was just one place I had to go. My desire to understand scarcity brain and find solutions for it led me on a two-year-long, forty-thousand-mile journey. Besides Baghdad, I traveled to the jungles of Bolivia, a monastery in the mountains of New Mexico, labs across the country, the backcountry of Montana, and even (sort of) outer space.

I wanted to understand our scarcity brain and the scarcity loop and meet innovators who have found a way out of it. These people understand the downsides of more. But they also realize there is something profoundly wrong with how we try to solve many of our modern problems.

If and when we realize that overconsumption is causing our problems, we're often told the solution is to simply shoot for less. Eat less food to lose weight. Buy less stuff or throw out excess stuff to spark joy. Spend less time on our screens to be happier. Do less work to avoid anxiety and burnout. Spend less money to fix our finances or overhaul our business.

But less, I'd uncover, comes with its own set of problems. And some robust new research shows that blindly aiming for less can change us for the worse. There are even times we should lean into excess.

The people I met on my journey are asking the more profound and challenging questions. But their efforts are revealing the answers that work. They've found that permanent change and lasting satisfaction lie in finding *enough*. Not too much. Not too little. Some have even flipped the scarcity loop to an "abundance loop," using the loop to do more of what helps us.

Colonel Mohammed Abdullah's office door swung open. The room fell silent. Out stepped the colonel himself.

"Where is the American?" he asked. Every head turned my way.

The Scarcity Loop

Long before Iraq, my journey started in my hometown of Las Vegas, a town that is to scarcity brain what Vatican City is to Catholicism. Few places better condense our modern ability to consume into one spot.

But of everything this city offers, nothing seems to trigger scarcity brain more than slot machines. Las Vegas wasn't built on winners. It was built on spinning reels encased in pinging, dinging, flashing cabinets that people play over and over and over—eventually to their detriment. Which explains why the machines are everywhere.

The casinos on the Las Vegas Strip are, unsurprisingly, vast labyrinths of them. But slot machines are also in our gas stations, grocery stores, bars, restaurants, and airport terminals. And people play these slot machines at all hours of the day, for hours at a time. They play at grocery stores at 6:00 a.m. The local diner at lunch or dinnertime. And I once saw a guy post up at a slot machine at a 7-Eleven and have a pizza delivered.

I asked the cashier if that was normal. "Are you kidding?" he said. "We have regulars."

But Las Vegas isn't the only place filled with people who are regularly irregular. Thirty-four states allow slot machines. And like Nevada, many of those states allow the machines beyond casinos—in all sorts of nooks and crannies of everyday life. And they're cash cows anywhere we put them.

The machines make more than $30 billion each year in the United States alone, or about $100 per American per year. It's more than we spend on movies, books, and music *combined*. And the figure rises about 10 percent every year.

I wanted to understand why. Why are the machines so uniquely captivating? Think: letting your perishables spoil because you got sucked into playing a slot machine called Kitty Glitter at the grocery store at 8:00 a.m. on a Tuesday.

I started by calling a handful of researchers who study problem gambling.

This was a dead end. Real dead.

The researchers accused casinos of using strange, near-subversive methods to lead us to gamble more. We've all probably heard of some of these. For example, one scientist told me casinos remove the clocks so we lose track of time while playing. Another PhD-holding antigambling researcher told me, "Casinos never want to have ninety-degree angles." The argument was that right angles supposedly force us to activate the rational, decision-making part of our brain. "Right angles throw you up against yourself as a decision-making person, which could slow down your rate of gambling on the machine," that researcher said. Yet another explained that casino slot machine music plays only in the pleasing key of C, which is said to relax us and, in turn, relax our wallets. All of those claims had even been reported widely in media outlets like the *Atlantic* and the *New York Times*.

But nothing added up.

A bit of common sense and a few visits to casinos proved that these assertions were either myths or standard business practices. For example, no, casinos don't display clocks on every wall. But neither does, say, your local Costco or Macy's or Home Depot. Most businesses don't hang clocks everywhere, I assume, because human beings wear wristwatches and own cell phones.

And when I visited some of Las Vegas's most profitable casinos, I found right angles everywhere. I mean everywhere. For hell's sake, slot machine screens are square. Some areas of casinos looked as if a cubist designed them.

I even contacted Peter Inouye, a slot machine audio composer. "I can confirm that I don't always use the key of C," he told me. He writes his slot jingles in all keys.

But most mysterious was one other thing that didn't add up. Most of those myths about the "subversive tricks" casinos use to get us to play slot machines had been floating around since at least the 1960s. Yet slot machine gaming was unpopular back then. Not only were slot machines not in gas stations or supermarkets; they were hardly on casino floors.

Then, around 1980, slot machines spread like a virus. They overwhelmed casino floors and went from earning little money to making up to 85 percent of casino revenues.

Perhaps it was thanks to all those right angles I saw in the casinos, but I had an aha moment of rationality. Instead of talking to people who want us all to *stop* gambling, I needed to speak to people who want us to *start* gambling. I had to do what always takes a journalist to the best answers. I had to follow the money.

This led me to a curious casino just fifteen minutes from my home.

It was the newest and most cutting-edge casino in town. It had the most entrancing slot machines, swankiest tables, cushiest hotel rooms, and finest restaurants the gambling industry could offer.

But here's the thing: Most casinos will do anything to get you in the door. At this one—not unlike a Baghdad police station—you're not welcome.

Black Fire Innovation rises like a giant Rubik's cube near the edge of the Mojave Desert in Las Vegas. The building is 110,000 square feet and four stories—all square windows and clean modern lines. It's a few minutes from the red rock and cacti of the open desert. But the wildland tames into pavement, and from there all roads lead to the heart of town.

From the expansive glass windows of Robert Rippee's office, I could see it: the Las Vegas Strip. The desert sun's unrelenting rays were beaming from its casino resorts. Cathedrals of consumption all dressed in neon and lining 4.2 miles of Las Vegas Boulevard.

Rippee was sitting at his desk, back straight. Advanced degrees hung from his office walls. But the man was not the type of dorky, PhD-wielding academic I expected. Clear designer glasses framed his face, and Buddhist prayer beads wrapped his wrist. He had a triathlete's build, with a flawlessly trimmed salt-and-pepper beard and tailored clothing.

This sophisticated air comes from the years he spent as an executive at one of the largest and most profitable casinos on the Las Vegas Strip. The job required analyzing human behavior data and then making decisions that altered the actions of millions of visitors. He pivoted into academia in his fifties to study the topic formally. Rippee wanted to understand, deeply, what makes us tick.

Half a decade before our meeting, Rippee was having lunch with an executive at Caesars, one of the world's largest casino companies. Caesars owns over fifty sprawling casino properties around the United States, where gamblers spend $10 billion a year. This executive was complaining about a particular problem.

Caesars was buying all sorts of new technologies it was told would improve its bottom line. For example, a slot machine with a new video-game-like feature that would compel people to play longer. Or an AI-powered data tracker that created detailed profiles of individual guests based on how they bet, ate, drank, and shopped and then generated cues that led them to spend more money in the casino.

But Caesars would have no clue if the new technology worked until it had spent millions of dollars putting it to use. It was the casino's own sort of gamble. And in this case, the house was losing.

So what if, Rippee proposed, Caesars partnered with the University of Nevada, Las Vegas? That's where Rippee and teams of other scientists study how technology affects human behavior in casinos. UNLV and its various research wings are like Harvard meets Cambridge meets Area 51 for shaping human decisions on the Las Vegas Strip and beyond.

They know what *works*. And it has little to do with geometry or musical notes.

What if they created a casino laboratory? What if they built a casino but used it entirely for research? Create a real casino, but fill it with a bunch of PhDs, brilliant tech minds, and study participants. It could be a place to test games and how people bet after obvious versus subliminal cues. Or investigate how the tiniest tweaks to a slot machine trigger us to do one thing or another. It could be an incubator where people with the next big idea could collaborate, find venture capital funding, and access the minds of brilliant scientists and industry insiders.

The result is this place. Black Fire Innovation. A sort of twilight zone of casinos.

"We're a giant lab that emulates a casino resort on the Las Vegas Strip," Rippee said. "A place where we can explore new technologies, behavioral changes, and more. We've emulated the entire integrated casino resort. From hotel rooms to food and beverage, to entertainment, to the gambling, and even the retail and signage."

Rippee told me this as we walked out of his office and into what was effectively a casino. Except there was no smoking and more scientific devices and people with advanced degrees.

"This area emulates a sports book," he said. Multiple betting windows and kiosks lined a wall. Dark gray overstuffed leather chairs faced a gigantic, credit-card-thin screen broadcasting a Twins-Yankees game. Flanking it were smaller screens across which scrolled the day's sports wagering lines. Astros -190 over the Rangers. Red Sox -125 against the Mariners. Mets even against the Cardinals.

We then walked toward a lineup of green-felt tables encased in dark hardwood and leather. Black and red rich leather chairs surrounded them. "These are traditional table games where we can test new games, technologies, and track behaviors," he said.

Rippee then pointed to a round machine surrounded by six chairs with a big screen rising from its middle. "We have electronic table games too," he said. At those, gamblers in the studies bet at individual screens as a virtual dealer on the larger center screen shuffles cards or spins a roulette wheel. You better believe the A.I. revolution hasn't been lost on the gaming industry.

Rippee then motioned to a hallway in the far corner of the room. It ended in two side-by-side doors, each with a key-card entry. "That leads to two hotel rooms," he said. Nearby was a big open kitchen. Beyond it was a cocktail bar. Then a coffee bar. In those places, researchers can test how every detail of the casino's rooms, food, and drink feed back into a guest's entire experience.

The tour continued. He pointed at walls consumed by video screens blasting test advertisements. Then at an interactive smart mirror designed to guide gamblers to critical points in the casino. Then at "a digital lounge for tech experimentation," he said. "And over there is an e-sports arena.

"So the overall idea," Rippee told me, "is that we can bring in a bunch of people and expose them to different scenarios. And then we can go back and measure their expectations and behavior. From there we can get some insight into how behavior changes as we advance technology.

"And this is all possible," he continued, "because we have seventy-three-plus companies who partner with us to provide cash or equipment." Caesars is the leading partner. But tech and gaming giants like Adobe, Intel, LG, Hewlett-Packard, Panasonic, Zoom, Boyd Gaming, and DraftKings are also financially vested.

Las Vegas casinos are not the Mob-run joints they once were. They're now living research labs and testing sites. Vast human behavior data banks. This is why, before the builders brought craps tables and roulette wheels into this place, they installed its brain: a supercomputing data mainframe. Rippee motioned to it.

The mainframe lived in an air-conditioned glass room and was the size of four refrigerators. It hummed, breathed. Multicolored wires emerged from it. They were all bundled together and crawling up the wall like veins, eventually disappearing into the ceiling.

Like the data mainframes in casinos across Las Vegas, the one here snakes its tentacles into every occurrence. What happens in Vegas no longer stays in Vegas. Every human action and its ensuing chain of reactions converges into the cloud, where it gets poked, prodded, and over-analyzed.

Rippee then led me to a final feature of this place. What I came here to understand. A workhorse for understanding the past, present, and future of human behavior. A metaphor for a new era of humanity where we're being pushed into behaviors we quickly repeat, and repeat, and repeat. For reasons we may not fully understand.

"Here are the slot machines," he said. They lined a wall. Their chrome cases glinted, and their screens all pulsed. "But I want you to speak to someone else about these," said Rippee. "His name is Daniel Sahl."

The people transforming humanity are no longer the oil barons of the early twentieth century. Or the Wall Street moguls of the 1980s.

They're people like Daniel Sahl. Mathy types who also "understand what'll engage people," as he put it. Sahl was wearing jeans and a track jacket that covered a T-shirt stamped with the Pizza Planet logo from the movie *Toy Story*. He could have used a haircut. He never made eye contact as he passed slot machines and card, craps, and roulette tables while walking toward me through his laboratory.

To understand why we are being pushed so deeply and rapidly into more, you need to understand the mechanics of the scarcity loop. And there's no better way to do that than by unraveling a peculiar shift that occurred in the slot machine industry around 1980. It amplified the scarcity loop and took it mainstream. Which brought me to the Center for Gaming Innovation.

The Center for Gaming Innovation is part of Black Fire Innovation. It started in 2013, when Sahl was a PhD student investigating how to apply video game theories to slot machines. It was a far-out theory at the time, but Sahl had a knack for dreaming up casino games that compelled people to play in rapid succession. Just two years later, UNLV gave Sahl the lab.

Games created in Sahl's lab have raked in untold amounts of money. Millions of dollars have made their way back to the lab. "We've got nearly thirty patents and counting. We've sold over thirty games to casinos around the world," Sahl told me after we sat at a semicircular card table in the lab. "And that money all goes back to the center and students. It's not often you can take a class, come up with an idea in it, and six months later get a check for an entire year's tuition."

And the lab's graduates don't just end up working for the world's biggest casinos, slot machine manufacturers, and mobile betting apps. They, too, follow the money. They're developing new behavior-modification technologies for U.S. military contractors, law enforcement, tech start-ups, and massive online retailers. Suppose you can design a game that is so engaging a person will play it hundreds of times in a row despite knowing they're likely to lose money. In that case, you can design products that compel people to repeat all sorts of other behaviors, too. These graduates are the future farmers of the scarcity mindset.

Through the 1970s, casino executives ignored slot machines and shoved them in corners. The 70s-era Las Vegas casino executive Charles Hirsch called slot machines "toys to occupy and amuse the friends and family of the casino's real customers—the card or dice players."

Card and dice games made ten times more money for casinos back then. Gamblers favored those games because they were loud, sexy, and exhilarating. Plus those "table games" allowed gamblers to experience the thrill of a win nearly half the time. Players had a 40 to 49 percent chance of winning any given game.

Slot machines, on the other hand, were boring. So few people played them. The machines were cumbersome, quiet, analog contraptions. Gamblers sat alone in silence pulling a handle and watching ugly steel reels spin. Then clunk, clunk, clunk—the reels would land. And the gambler would probably lose.

That was the biggest problem. Gamblers could bet only a single row of symbols each game. So to win, the gambler needed the right symbols to line up perfectly in the row down the middle. This happened rarely: only 3 percent of all slot machine plays won.

Common sense dictates we will quickly stop doing something if it gets us nothing. A century of psychological research also backs this up. For example, if we turn the key to our car and the engine doesn't start, we might turn the key a handful more times. But if nothing happens, we're not going to keep turning and turning the key. We'll give up and open the hood or call a tow truck.

Psychologists call this stopping of unrewarding behavior "extinction."

It's apparent in every animal we've studied. And when it came to causing extinction, the slot machines of the era were like the comet that killed the dinosaurs. They were so good at causing extinction that many didn't even have chairs. Few people played them long enough to need to sit.

Then came a man named Si Redd. This was around 1980. Redd, born in 1911, was the son of a dirt-poor Mississippi sharecropper. Growing up in the Great Depression had altered his psyche. It branded him with an unrelenting drive and energy that he focused entirely on getting rich. At just eighteen, he began building an empire of pinball machines and juke-boxes across the South and Northeast. Redd was so successful that by the end of the 1950s the Mafia made him a raw deal: sell us your business or we'll kill you.

So he took his talents to Las Vegas. Redd rolled around Vegas in maroon polyester suits, saucer-sized sunglasses, and bolo ties studded with golf-ball-sized chunks of turquoise. Cowboy meets Rat Pack. Old-school Vegas.

In the late 1970s, Redd noticed that the new Atari video game system was holding kids' attention for hours. Which to him was silly—because when these kids won a video game, they got nothing tangible for it. But it gave him an idea.

Redd understood the psychology of extinction. He knew that experiencing too many losses in a row is no fun. Gambling is far more exciting when we're winning—even if those wins are small.

He wondered, could he digitize slot machines? Instead of having those clunky physical analog reels that offered just one row of symbols to bet on and terrible odds, Redd began manufacturing slot machines with screens. This meant that when gamblers played the game, it would trigger a computer and the "reels" would appear spinning on a screen.

Screens opened up a world of betting and winning possibilities. Redd had the machines programmed so players could bet on more than just one row, or line, of symbols per game. Some machines allowed gamblers to bet on a hundred lines of symbols in a single game. Imagine all kinds of wonkily shaped lines running across a five-by-five grid of symbols. Straight lines, downward-angled lines, upward-angled lines, V-shaped lines, M-shaped lines, and on and on. And *any* of them could win.

By betting even a penny or nickel on each of the ten, twenty, forty, or even hundred different lines every game, gamblers were more likely to win *something* on a line or two. The odds of winning on any single slot machine play spiked to around 45 percent.

It could be a big win. But far more often it was less money than the original bet. For example, a gambler might bet $1 and "win" 50 cents.

Framing this as a "win" might seem odd. Totally dumb, even. But science going back to the 1950s confirms Redd knew something fundamental about human behavior. He understood that the human brain doesn't experience that result as *losing* 50 cents. It tends to ignore the dollar invested and perceives this as *winning* 50 cents. Casinos call winning less than we bet "losses disguised as wins."

And researchers in Norway recently discovered that our brains respond to these "losses disguised as wins" as small wins rather than small losses. They lead us to play longer and spend more money because they maintain hope, suspense, and excitement.

Once he'd leveraged this brain quirk, Redd strengthened it by stealing a play from basic psychology textbooks. He made the machines near epilepsy inducing. Digital machines allowed Redd to add loud and upbeat sounds, bright flashing lights, and entertaining on-screen graphics. He programmed the machines to emit those exciting noises and lights and graphics for both true "wins" *and* "losses disguised as wins." Psychologists call what happened next conditioning. Just as Pavlov's dogs salivated at the sound of a bell, humans began associating the machine's fantastic reactions with not only big, true wins but also losses disguised as wins.

To understand this shift, let's say we were to play both an old machine and one of Redd's machines. Pretend we have $10 and bet $1 each game. Here's what it would be like.

The Old Machines

Game play: Lose, lose, lose, lose, lose, lose, lose, lose, win $2, lose, lose, lose, lose.

Net result: Lose $10.

Total time playing: One minute.

Excitement and likelihood of playing again: We'd rather have a root canal.

Redd's Machines

Game play: Lose, win $.50, win $.80, win $1.50, lose, win $.40, lose, win $.80, win $.25, lose, lose, win $4.00, lose, win $.50, win $6.00, lose, lose, win $.20, lose, lose, and so on until . . .

Net result: Lose $10.

Total time playing: Fifteen minutes.

Excitement and likelihood of playing again: Can I borrow $10?

Slots shifted from a fast, dull burn into a long, slow, captivating, fun-as-hell smolder. One that kept us coming back. Sometimes we'd even walk away with more money than we started with.

To non-gamers, playing through losses disguised as wins can seem irrational. But it's textbook human behavior. Let's return to the example of our crappy car. Pretend we turn the key to our car, and nothing happens. So we turn the key again. Nothing. But when we turn the key a third time, the engine cranks and putters for a second, making sounds as if it were going to turn on. Then it goes silent and doesn't come to life. With this "loss disguised as a win" we will, of course, immediately turn the key again. We will sit there turning the key so long as the engine gives us signs of life. It's the stringing together of too many failed attempts that prompts us to open the hood or call the tow truck. A finicky engine keeps our attention longer than a dead one.

By making "wins" frequent and adding dazzling lights and sounds and graphics, Redd solved the machines' boredom problem.

But he had one more problem to solve. People didn't *realize* his new slot machines were so fun. People still thought slot machines were dull one-armed bandits. Redd needed to entice the masses.

Screens and digitization allowed him to solve that problem in the same way we solve many issues. He threw money at it.

The jackpot odds of the old machines were determined by the number of symbols engineers could fit on a reel. Those jackpots were nice— maybe $500 to $1,000. But any schlub could more easily win that much money if he had $100 and got lucky on a few games of blackjack or roulette.

Digital machines could be programmed with any odds Redd wanted. Say, 1 in 250,000 odds for a jackpot. This allowed him to offer bigger payouts. The average jackpots of the new machines swelled into the five- and six-figure range. Redd even connected machines from across the entire state of Nevada and pooled their money to offer multimillion-dollar jackpots. He called these "wide-area progressive jackpots." They were like the Powerball lottery of slot machines.

This opportunity to win a life-changing jackpot by betting just a dollar or two enticed the masses to try slot machines. Sort of like how when the Powerball jackpot reaches nine figures, your local gas station has a line of people out the door all scrambling to buy lottery tickets.

More people tried slot machines. And once they did, they had fun and kept playing. Slot players knew they'd win soon. But how soon? And how much would they win? Would their dollar bet get them 40 cents? Or would it win them $40 million, as it did for a twenty-five-year-old software developer who won as much on one of Redd's wide-area progressive slots in a Las Vegas casino in 2003?

Redd didn't stop there. He was always looking for the next tweak that got more people playing more. Those cumbersome handles on slot machines could make playing feel like manual labor. So he got rid of them and added Spin buttons. The man didn't care about your tired arm, but he did care about his bottom line. Spin buttons let gamblers play the next game quicker. The average player went from playing three hundred games an hour to nine hundred.

Redd's machines took off. People clamored to play them. Casino executives increased the number of slot machines on their gaming floors fivefold. They redesigned casino floors to put slot machines front and center. Executives even had to order thousands of chairs because slot players were gaming for hours and needed to sit.

Everyone got rich. Except you, me, and anyone else playing the machines.

Redd's slot machine revolution was one of those fundamental shifts that occur once in a century. Like when Netflix began allowing customers to stream their movies instead of mailing them back and forth in the form of DVDs, altering how we consume TV and video forever. Or when Amazon thought to sell more than just books and changed how we shop.

Redd had by instinct tapped into a powerful quirk in the human mind. The behaviors we do in rapid succession—from gambling to overeating to overbuying to binge-watching to binge drinking and so much more— are powered by a "scarcity loop." It has three parts.

Opportunity—> Unpredictable Rewards—> Quick Repeatability

This loop is the ultimate trigger of the scarcity mindset. What Redd uncovered is near-compulsive repeat consumption. No matter the behavior, the more we have the opportunity and desire to quickly repeat it, the greater its effect on us.

His hunch for how the scarcity mindset works and what sets it off increased slot machines' revenues tenfold and rendered table games a sideshow.

Right now, new generations of behavior engineers like Sahl are refining Redd's work. They're using a century's worth of psychological research and casino data to optimize the scarcity loop to make games even more compelling. All to sharpen the slight house edge.

Once Sahl had covered the success of the lab, he wanted to talk about math. This was his comfort zone.

"Every casino game returns an average of 87 to 98 cents of every dollar a gambler wagers," he told me. This gives casinos about 7 percent margins. That's a relatively low amount. Profits rely on volume. Getting lots of people to play for a long time.

Enter the scarcity loop. Sahl began describing to me why slot machine gambling is so compelling. "So when the reels are spinning, what happens for the gambler is . . . ," he said while staring at the table. Then he trailed off.

"Here, let me just show you," Sahl said. He walked us to a machine. The game was called Scarab. Its theme centered on ancient Egypt. The screen displayed decorated symbols resembling ancient Egyptian hieroglyphs. Sahl began jabbing at the machine's Spin button.

The reels on the machine's screen rolled while it projected a barrage of sounds and shimmering gold highlights, focusing our attention on the valuable symbols. Sahl then began a primer on slot machines and their powerful use of the scarcity loop.

1. Opportunity

The first part of the scarcity loop is opportunity. An opportunity to get something of value that improves our life.

But this opportunity comes with risk. We might get something valuable. For example, money, possessions, food, or even status. But we also might not—or we could even lose it.

"Gambling is so compelling because there is quantifiable risk associated with the reward," Sahl told me. "It's not just possible to win. It's also possible to lose something that is actually valuable across society. Money is tangible. The risks and opportunities are clear."

Research conducted at Columbia University backs his claim. The scientists discovered that the more an event could lead to a clear reward or loss—for example, winning a jackpot or learning the results of a medical diagnosis—the more we enter a trancelike state of fixation as we wait for the outcome.

2. Unpredictable Rewards

The second phase of the scarcity loop is unpredictable rewards. The rewards of everyday actions are predictable. If we have an itch and we scratch the itch, we know the itch will be relieved. The reward, the itch going away, is predictable.

But decades of research show that these predictable rewards can be dull.

Unpredictable rewards, on the other hand, are not. If we know we'll receive a reward but aren't sure when, we get sucked in. We experience a sort of exciting, suspenseful anxiety as we wait to see whether this occasion will deliver the good stuff. Our brains hone in on unpredictability. They naturally suppress systems that take in other information, and we fixate on whatever is unpredictable. One study found that unpredictable rewards "tap into fundamental aspects of human cognition and emotion."

Between spins, Sahl was explaining the details on this second phase of the scarcity loop. His attention would shift to me just often enough to remind me he wasn't entirely ignoring me in favor of playing the game. "The key to gambling is that you're anticipating a reward," he said. "You know you'll probably get a reward eventually, but you don't know when or exactly what it will be."

Sahl hit the Spin button again. "With this spin," he said as the wheels rolled and began to set, "are we going to be disappointed? Or are we going to be happy? And will we be just kind of happy or will we be really, really happy? That's what's exciting. The result of a spin could result in nothing. Or it could be life changing."

And there are, Sahl told me, other ways he dials up the anticipation of unpredictable rewards. He masterfully engineers how a potential win unfolds.

Recall that everyday behaviors occur the same way every time, like our example of scratching an itch. Slot machines change that dynamic.

He hit the Scarab machine's Spin button. Its five reels began to roll. Reels one, two, and three all landed showing symbols that indicated the beginnings of a winning combination. "So right now, we'll get a big win if the final two reels stop on the right symbols," he said. "But as a game designer, I don't want those final two reels to go through the standard motions and stop quickly, at the same rate they would if the first three reels were losers. I want to extend this potentially winning experience."

The two remaining reels kept spinning much longer than usual. Shimmering white light outlined the fourth reel, lighting up our faces as it spun and emitted a fantastic, mysterious song on theme with ancient Egypt.

Truth be told, I was captivated. My face surely looked as if I were getting beamed up by some martian mother ship. I was only half paying attention as Sahl said something like "so what is happening now is that this is dialing up the suspense and holding us in the moment longer and more intensely."

These moments when we wait to learn the outcome of an unpredictable reward are, in fact, so exciting that Stanford neuroscientists discovered that they become rewarding in and of themselves. The brain's excitement and reward circuitry react strongest during these moments where we're waiting to find out if we got the reward.

The fourth reel landed. We were still on track for a win. The music continued and the light shifted to circle the fifth reel, which spun and spun.

"Damn it!" I shouted, surprising myself as the final reel landed. We didn't get the symbol that would have brought us a big win.

"Nope, we didn't win," said Sahl, nonreactive to my outburst. "You realize we're not playing with real money, right?"

I just smiled and shrugged. A win's a win. Sahl continued: "See . . . we *almost* won, but we lost." The engine, despite all noises to the contrary, sputtered out.

"Casinos call what just happened a near miss," Sahl explained. "And near misses are key. They appear in all games, but they're critical in slot machines. They provide entertainment, excitement, and stimulation and compel people to play again quickly. But in slot machines, losses are mathematically far more likely to occur than wins, so you have fun, but the house doesn't have to pay you anything."

The idea that a near miss would lead us to play again immediately is another one of those things that seems weird but isn't. Psychologists have been observing it for decades. They can even plot it on graphs. Let's say we do something and expect something to happen. If that "something" doesn't happen, we immediately repeat the behavior. Fast and hard. For example, we hit an elevator button. If the button doesn't light up, we'll quickly jab at the button a bunch of times in a row. Another example: A child says "Mom" and gets ignored. He'll follow that up with "Mom. Mom. MOM, MOM, MOM, MOM . . ."

Slot machine engineers lean on elegant math that results in constant near misses, causing people to play again faster. Teams of scientists in Israel and Canada discovered that frequent near misses—the same amount we see in modern slot machines—lead people to gamble about 33 percent longer. The scientists wrote that near misses "invigorate play," because our brains register these near misses similarly to wins. As Sahl said, the near misses become fun and rewarding in and of themselves, even though we lost.

Now that our game was over, the Scarab machine was near lifeless. "See how the machine went quiet when the last reel didn't line up for a win?" Sahl asked. "That's on purpose. We lost. Next bet. Why would we want to draw attention to a disappointing outcome?

"But if we were to have won," said Sahl, "that experience would be dragged out even more. If you have some symbol combination that pays you back 150 times what you bet, I want to take as much time to show you those lights and noises as I can and project special graphics on the screen. I want to tell you that in a two-minute story." It all becomes spectacular beyond the money we won, making hitting that Spin button again even more enticing.

3. Quick Repeatability

The third phase of the scarcity loop is quick repeatability.

Most everyday behaviors have a clear beginning and end, and we don't immediately repeat them.

We have an itch, we scratch the itch, and the itch goes away. Or our hands are dirty, we wash them, and our hands become clean. We have no reason to scratch or wash our hands seconds later. In fact, quickly repeating the behavior can punish us; we'd scratch off our skin or wash our hands raw. (And, if we continue, we call that behavior compulsive.)

Scarcity loops, on the other hand, are immediately repeatable. We see opportunity, receive rewards sometimes, and then do it all over again. As much as we'd like.

Sahl laid out a bunch of technical material around the "repetition interval" and "reinforcement." But the takeaway is that quicker repeatability is better.

"People today do make decisions that involve opportunity and unpredictable rewards, like our diet choices, financial planning, or buying a house. But we rarely get resolutions quickly," he said. "It might take us ten, thirty, or even fifty years to learn whether our house is a gold mine or money pit or if our diet led to health or disease."

Gambling is different, said Sahl. Slot machines allow us to learn whether an opportunity paid off within seconds and incentivize us to play the game again immediately. Research shows the faster we can repeat a behavior, the more likely we are to repeat. Speed kills. Redd's hunch to shorten the time between games paid off.

And that's it. Those are the three conditions for a behavior to fall into a scarcity loop: opportunity, unpredictable rewards, and quick repeatability.

But how do we get out? A person stuck in a scarcity loop stops for only three reasons, all of which jam a stick in the spokes of the loop.

First, the opportunity could go away. For gamblers, this could be from running out of money or, in the rarer occasion, making enough that they feel satisfied to stop.

Second, the rewards could stop trickling in. For a gambler, this is stringing together too many pure losses in a row. Which explains why so few people played early slot machines.

Third, the repetition could stop being quick. This is rarer for gamblers, but it could be that the gambler gets physically tired or the Spin button starts jamming.

For Sahl and me, it was the first option. We ran out of the fake money he'd loaded onto the machine. "Well," he said, "that was fun. Want to play another machine?"

All casino games can push us into a scarcity loop. Blackjack and craps, for example, have opportunity, unpredictable rewards, and quick repeatability. It's why they were the casino workhorses before Si Redd came along.

But slot machines won because they amplified the loop. More and faster and stronger. For example, the average blackjack player can play 60 to 105 hands an hour. But in that same time, a slot machine player averages anywhere from 600 to 1,000 games. Let's say she's betting forty lines a game, which is a typical number. That amounts to anywhere from twenty-four thousand to forty thousand individual wagers—with a fantastic range of unpredictable rewards.

"Okay, interesting," I told Sahl. "But why do people play these machines when they know they probably won't get their money back?"

"You're right that everyone knows the house always wins," said Sahl. "But you're asking the wrong question. You're assuming people play only to win. Gambling allows us to experience risks and thrills, and that's fun."

Research indeed shows that most people don't gamble with enough money to really affect their finances. Rather, gambling is a hobby.

From an economic perspective, any hobby that costs time or money is a losing venture. The gambler could just as much ask, "Why pay $50 for a concert ticket when you know you're not going to get the money back? Why pay $100 to play a round of golf when you're not getting the money back?"

If a gambler loses money playing a slot machine but still had fun, they have, in a way, won. If they happen to leave the casino with more money, they've double-won.

See, that's the thing: Falling into the scarcity loop can be fun. The combination of opportunity, unpredictable rewards, and quick repeatability provides the structure for the ultimate game. But if we play it too often and escape into it for reasons beyond fun, the problems can pile up, said Sahl.

"Slot machine gaming relies on an extremely powerful system," he said. "Extremely powerful. There are definitely some people who gamble too much of their income, and I feel terrible for those people."

And so did Redd. When problem gambling rates spiked thanks to Redd's new machines, many people accused him of fueling addiction. In 2001, two years before Redd died, a *Las Vegas Sun* reporter asked him about this. "Of course it hurts me when such things are said, I guess because it is kind of the truth. I never intended it to become that way. And I could have never dreamed of how successful the machines would become." About 1 to 2 percent of the general population now qualify as compulsive gamblers.

I began thinking about some of my own behaviors in the past and present. Alcoholism I've been in recovery from since 2014. Mindless eating. Endless scrolling. Amazon Prime shopping to distract myself from doing deeper and tougher work. Could some or all of that be connected to my own scarcity brain and tendency to get sucked into the scarcity loop?

Before I left, I had one last question for Sahl. "I get that gambling is fun," I said. "But why is it so engaging in the first place? What's the deeper reason for why?"

Sahl just shrugged. I'd have to speak to someone else about that.

How The Scarcity Loop Hooks Us

Thomas Zentall began studying psychology in the 1960s. While many other twentysomethings in the Bay Area were dropping acid or staging protests, he was pursuing a PhD at the University of California, Berkeley. He graduated in 1968 as occurrences like the Gulf of Tonkin incident and the CIA's MK-Ultra program were revealing to the nation that larger governmental and corporate forces are often pushing and pulling us in directions we aren't always aware of.

Zentall is now in his eighties. He officially retired in 2019. But he made a deal with the University of Kentucky, where he conducted pioneering psychological research for five decades. He'd teach a graduate-level class for free if they'd let him keep his office and lab. He still spends sixty hours a week in the lab.

I wanted to understand why the scarcity loop is so captivating. Zentall would be the person to ask.

After earning his PhD, Zentall began following up on counterintuitive research conducted by the famed psychologist B. F. Skinner in the 1940s. Skinner started uncovering the scarcity loop's power while working with lab rats.

Skinner typically gave the rats their favorite rat treat each time they'd hit a lever. But he'd run low on treats. Apparently, Skinner was feeling rather lazy because he didn't make more treats. He saved remaining treats by giving the rats their treats at unpredictable intervals when they hit the lever. Skinner assumed the rats would get bored or bummed out. After all, the rats received far fewer rewards for pressing the lever.

He was wrong. The opposite happened: The rats developed a near-OCD-like affection for hitting that handle. It sucked them into the loop of opportunity, unpredictable rewards, and quick repeatability.

Skinner's observation fascinated Zentall, and he agreed to meet me over videoconference. His hair was white and face pale and soft from decades in his fluorescent-lit laboratory.

"I started conducting animal research because I was interested in the degree to which human learning and behavior could be explained in terms of simple processes we see in animals," Zentall told me. "The same psychological mechanisms that affect animals affect humans." He's since discovered that it's quite easy to turn pigeons into degenerate gamblers.

He conducted a study where pigeons chose between two games. In game one, the pigeons received fifteen units of their favorite food every other time they pecked a light. So 50 percent of their pecks led to food. Like, peck: no food. Peck: food! Peck: no food. Peck: food! And so on and so forth.

In game two, they got food about every fifth peck. So only 20 percent of their pecks led to food. But this game had two catches.

The first catch of game two is that the win was larger. When the pigeons won, they got twenty instead of fifteen units of the same delicious food.

The second catch is that the wins were unpredictable. So this game was more like this:

Peck: no food
Peck: no food
Peck: no food
Peck: So much food!
Peck: no food

But the next round of five might be something like this:

Peck: no food
Peck: so much food!
Peck: no food
Peck: no food
Peck: no food

You get the point.

Zentall explained that the math shows it makes far more sense to play the first game. If a pigeon plays game one a hundred times, it will end up with a total of 750 units of food. But game two, even though a win is worth more, would get the bird only 400 total units of food. There is, in fact, an entire scientific idea called "optimal foraging theory" that says animals do whatever they can to get the most food for the least amount of effort. So that theory, along with common sense, tells us that any pigeon that plays the low-odds game is a total dipshit.

We know that every society—whether of humans, pigeons, rats, or whatever—includes a handful of dipshits. But when Zentall ran this exact experiment, "within just a couple of sessions, the pigeons start preferring the low odds side that gets them less food," he said. Exactly 96.9 percent of pigeons chose game two.

Zentall's eyes were wide with a look that screamed "Can you believe that!" as he told me the same phenomenon is seen in cockroaches, monkeys, rats, mice, other birds, and more. Laboratory animals consistently play the gambling game even though it nets them fewer resources. Some will choose the gambling game even if the predictable game gives them more than 700 percent more food. "Seven hundred percent," Zentall repeated.

He continued: "We consistently find that the value or reward of a behavior depends in part on how often you think you'll get the reward. If you are almost certain the reward will happen, it's nice. But if you are unsure the reward will happen, then you're very excited when it happens. So much so that you'll make suboptimal decisions. And we see it in humans all the time. In all different areas."

"But why?" I asked. "If it's not optimal, why do this?"

Zentall laughed. He paused and tilted his chin toward the ceiling as he took in a big breath of air. "Ahhhh," he said. "I get that question a lot."

The answer lies in the scarcity loop. Our ingrained attraction to it developed thanks to a consequential gamble that all species make every day. One that we humans used to make daily but no longer do. Finding food.

Humans evolved to cover great distances looking for something to eat. In the past, we'd typically walk or run between five and thirteen miles a day while hunting and gathering. This was effectively the ancient form of picking up groceries. Except we often had no clue where food was. So we'd search and search the land. These ancient searches were like playing a modern slot machine.

"When our ancestors didn't find enough food in three, four, or five places, it didn't stop them," Zentall explained. "They'd just keep searching." Finding food is our original and most important opportunity to survive and better our life. If we didn't keep searching and searching—pulling and re-pulling the metaphorical slot handle—we'd starve and die a prolonged and excruciating death.

We'd travel to one area we thought would have food (pulls handle, and reels spin and land). It didn't have anything. So we'd go to another location (pulls handle, and reels spin and land). Nothing. Then another area (pulls handle, and reels spin and land). Zip. Then the next place (pulls handle, and reels spin and begin lining up . . .).

Jackpot. This place had all kinds of food (the machine lights up, goes ballistic, and starts spitting out money).

Like playing a slot machine, our searches kept us in suspense with unpredictable rewards. We knew we'd probably find food eventually. But when? And how much food?

As Daniel Sahl put it, "Gambling isn't when you place the bet or learn whether or not you won. Gambling is when the cards or slot reels or dice are falling." Or, in our past, when we, on foot and hungry, searched the land for food.

As we searched, we experienced near misses and losses disguised as wins, those times we come close to winning but don't or win less than we bet. Let's say we spotted a distant berry bush. "The probability of finding food then changes," Zentall explained.

Was it a lone bush devoid of fruit, like a near miss? Or did it have just a bit of fruit, not enough to replace the energy we burned walking to it, like a loss disguised as a win? Or was it a true win? Was this bush packed with berries and on the periphery of an entire grove of fruitful berry bushes?

If we were hunting, was that animal in the distance skinny and small or big and fat? And was the animal alone or with an entire herd? (Slot reels line up indicating a win, and we wait in suspense to learn just how big of a win.) We would experience near misses if the animals got away.

And, of course, we'd repeat this behavior for much of the day, every day. Evolution drilled our attraction to the scarcity loop into our head. It's precisely why our brains reinforce falling into the scarcity loop. It was an opportunity to survive and better our lives, with unpredictable rewards, that we quickly repeated.

But this ancient game didn't apply just to food, said Zentall. It applied to acquiring anything that gave us an opportunity to improve our lives. This could be gaining possessions or other resources, information, social status, or whatever else made us feel good and live another day.

For us to survive, scarcity brain needed to develop systems that pushed us into the loop. Which leads us to dopamine, the brain chemical that is as famous as it is misunderstood.

By 1990, scientists could see that dopamine is linked to all the fun stuff humans do. They found that sex, drugs, and gambling all cause our brain to shoot out dopamine. "So dopamine was considered the pleasure neurotransmitter," said Kent Berridge, a neuroscientist at the University of Michigan. The theory was that these acts were just a means of chasing a dopamine high.

"This idea still persists today," Berridge told me. Dopamine has since achieved mythic status in pop psychology. We're told we're a nation of dopamine-addled zombies, all out jonesing for our next hit of the stuff. We hear of people who do all kinds of counterproductive and depraved acts for the dopamine high.

For example, one article in *Forbes* titled "Addicted to Bang" blamed dopamine for people's enjoyment of firearms. Or there was one from NBC News titled "Why QAnon Followers Are Like Addicts." Apparently, dopamine, this story explained, caused people to believe that "a cabal of Satanic, cannibalistic pedophiles operate a global child sex trafficking ring that conspired against former U.S. President Donald Trump" (as QAnon has been described).

We're even told that a "dopamine fast" will solve our problems. But that's basically just a poppy-neuroscience-babble way of saying, "Stop doing stuff you don't want to do anymore." Which is all to say dopamine has become a bit of a patsy for every bad human behavior ever.

But in reality, dopamine isn't the "pleasure chemical." And it doesn't *make* us do or believe anything. "It makes us more likely to pursue reward, even in situations where that reward is maladaptive," Berridge told me. It's also a versatile, upstanding citizen. It's a neurotransmitter, which means that it transmits information between nerve cells in our brain and body.

It also acts as what's called a paracrine messenger. This means that it works like a construction contractor, commanding all sorts of other processes to happen. It regulates movement. Parkinson's disease, for example, is caused by specific brain cells not producing dopamine. It helps us urinate. It influences how food moves through our digestive system. It helps control our white blood cell production, which allows us to combat invading viruses and tumors that could kill us. It regulates insulin, which regulates our metabolism and how we burn the food we eat.

If we were to truly fast from dopamine, we'd die fast. From Parkinson's, a full bladder, a virus, or any other vile malady.

When it comes to behavior, dopamine helps us associate certain conditions in our environment, or cues, with getting a reward. Once we know something is pleasurable or rewarding, dopamine is primarily released when we're *pursuing* and *anticipating* receiving that pleasurable thing, not when we're *actually* receiving the pleasurable thing.

Hence, most popular discussions about dopamine are neurosciencey ways of explaining a rather basic and observable formula for how all animals learn and crave and, in turn, behave. The formula is this: *We do a thing. If the thing was rewarding, we're now more likely to do the thing again in similar conditions.*

That formula drives everyday behaviors. For example, let's say we arrive home stressed from work and have a glass of wine. If that glass of wine relieves our stress, it's as if our brain goes, "Aha! That was good. Remember that."

On future days when we arrive home stressed, our brain will likely release dopamine to create a craving for a glass of wine. And this process is deeply embedded. Take recovering drug addicts. They can be clean for years. But if they find themselves back near a place where they used to score, their brain will still often pump out dopamine, causing them to crave their drug of choice.

But the scarcity loop is so powerful because it leverages a hiccup in the learning and behavior formula. As the work of Skinner and Zentall shows, all animals, including humans, want a reward *infinitely* more if we aren't sure we'll get it. If it's received unpredictably. Unpredictability makes us obsessive and far more likely to quickly repeat the behavior.

"The expectation of *possibly* receiving a reward really excites the dopamine system far more," said Zentall. "Dopamine peaks when we don't know if we'll get the reward." Unpredictable rewards suck us into a vortex of suspense. Near misses and losses disguised as wins stoke the system to incentivize quick repetition.

So the tweak to the basic learning and behavior formula is this: *We do a thing. But we're not sure when we'll get the rewarding thing or just how rewarding it'll be. This makes us really, really want the rewarding thing. So we're likely to keep trying and trying and trying and trying for the rewarding thing.*

Sure things are boring. Unsure things captivate us and make us likely to repeat again and again.

Consider the slot machine. It would be nice if it paid the same amount every game. But playing it wouldn't be exciting. It would be called a job. This is essentially what salaried or hourly wage jobs are. We repeat a behavior, and an employer gives us a predictable reward.

So the loop is an ancient game that developed to keep us alive. It compelled us to persist in the face of uncertainty and do it all over again. Quitters die.

Today, survival is easier and we don't need to spend our days searching for resources. But the scarcity loop still captivates our brain.

Where The Scarcity Loop Lives

When I started looking for the loop, I began to see it everywhere. "By the way, this system is not just fundamental in slot machines," Sahl had told me. "It's now becoming fundamental in the design of a lot of other products too. It's everywhere," embedded in many of the institutions, technologies, and experiences impacting our lives.

Opportunity—> Unpredictable Rewards—> Quick Repeatability

That elegant, three-part system seemed to be woven into so many of our everyday behaviors. And it was unmatched at altering our actions. I could now see it was leading me to spend more money, capturing and holding my time and attention, and prompting me to repeat habits I wanted to limit.

Take how it worked me over during the pandemic. I ended lockdowns with ten more pounds on my frame, more hours on my screentime, less money in my bank account, and all sorts of unnecessary possessions. That was thanks to working from my kitchen counter and eating crunchy, calorie-dense food when the job became frustrating; from television and diving down insane, bottomless internet rabbit holes of information and social media; from betting on obscure sports or buying crap online for something stimulating to do; and so much more.

Waking up to what was happening was like the feeling you might have after you realize you've forgotten some important work meeting. It's a sinking panic followed by frantic and fruitless attempts to smooth over irreversible and embarrassing ineptitude.

And yet, at its core, the loop had nothing to do with the pandemic— it's always been part of our lives, affecting us in the best and worst of times. The only thing I could do from there is to become conscious of the loop and how it was affecting me and those around me. I began sussing it out.

SOCIAL MEDIA

The loop obviously makes social media "work." Posting offers us an opportunity to enhance our social standing and each notification feels like the unpredictability of spinning reels. Did we get a like or comment or DM? How many? And did this deliver news of social acceptance, such as likes or positive comments? Or rejection? Minimal likes or some snarky comment.

Scrolling the feed also tosses us into the loop. It's a search for something that might make us happy, sad, annoyed, outraged, envious, or surprised. Infinite scroll allows for quick and endless repetition.

EMAIL

The loop lives in our email, which has become mandatory for our personal and professional lives. Refreshing our inbox or receiving a notification creates an unpredictable moment of suspense where we may receive good, bad, or just OK news.

SHOPPING

The quick, compulsive repeatability offered by Amazon Prime is one thing. But it's nothing compared to new apps like Temu, a casino-like online store that sells items direct to the consumer from factories in China. It's Amazon meets TikTok meets slot machine. "To engage with Temu is to be cornered in conversation with an AI-powered salesperson who is ushering you past endless tables of assorted goods to sell, *right now*, with escalating special offers, chained promotions, exclusive limited-time discounts, and lots and lots of free stuff," explained *New York* magazine.

The scarcity loop is also in our ads. Advertising analysts at Adweek recently reported that more digital advertisers are using casino-like unpredictability to drive sales. Think spinning a virtual roulette wheel to determine the size of a discount. Research from Deloitte found that ads embedded with unpredictable rewards increased customer engagement by 40 percent and yielded a sevenfold conversion rate.

PERSONAL FINANCE

The scarcity loop is now in our personal finance apps. Popular new stock-trading apps like Robinhood are leveraging even more unpredictable rewards and quicker repeatability. Robinhood doesn't charge its users to make trades. Instead, it bakes in trading fees through a unique and controversial practice called payment order flow. Many users quickly repeat trades and make hundreds of daily transactions, hoping to accumulate small unpredictable wins.

The executive director of the National Council on Problem Gambling spent a few minutes on the app and told NBC News, "A lot of this [the tactics Robinhood uses] is directly taken from the user experience of casinos: It encourages immediacy and frequent engagement."

Because of this, the app recently quadrupled in value and added thirteen million users, bringing its total to around twenty-three million. Now copycats like Webull and TradeStation have popped up. Even old standby trading platforms like E*TRADE and Charles Schwab now use tactics similar to Robinhood's. The finance world, the analysts believe, will continue seeking novel ways to exploit the scarcity loop.

MOBILE GAMBLING

We no longer have the barrier and pause of having to travel to a casino to play slot machines or bet on sports. The casino is in our pocket. "The biggest innovation in gambling is mobile," said Sahl. "Mobile gaming has been a game changer. It's totally expanded gaming." Sports betting has dialed up quick repeatability by letting us wager on down-to-the-second in-game occurrences, like whether a team will score on the next play.

Thanks to this new accessibility and ease of quick repetition, mobile gambling recently grew 270 percent in a single quarter and is worth fourteen figures in the U.S. alone.

TELEVISION

The scarcity loop is in the videos we watch.

Robert Sweeney, the developer who built Netflix's autoplay feature, said, "Autoplay massively increased hours watched. It [led to] by far the biggest increase in the hours watched . . . of any feature we ever tested." He continued, "Netflix wants you to spend more hours watching Netflix, and the product team is scientifically engineering the product to make it more addictive."

The loop is particularly pernicious on YouTube. The techno-sociologist

Zeynep Tufekci discovered that YouTube's autoplay algorithms lead us into successively more extreme and polarizing videos. Extreme content captivates you, me, and everyone we know because human attention naturally gravitates to information that implies danger or drama. It's an ancient survival mechanism being leveraged to capture our attention.

HEALTH

The scarcity loop is changing our health habits. Consider the WHOOP activity tracker and others like it, which use untraditional methods to leverage the scarcity loop.

The devices contain no concrete metrics that are predictable and easy to modify, like step counts. Instead, they lean into the suspense of unpredictable rewards by giving users a different daily "recovery score" and "strain score." What we do across the day alters these abstract scores in unpredictable ways.

Unpredictability leads users to check and recheck their scores throughout the day, every day.

For the dedicated, the recovery score determines the course of the day: whether a person will rest or go to the gym, what they'll do there, and how hard they'll do it. This health scarcity loop is altering behaviors in strange ways. One of my friends, a leading doctor of physical therapy, told me he has clients who come to him saying they feel great but refuse to exercise too intensely because their device tells them their recovery score is low. "We're making health decisions off of very questionable assumptions and data, basing important decisions around some abstract gamified number," he said. "It's madness."

Given WHOOP's success, other health trackers like Oura Ring, Fitbit, Apple Watch, and Lumen have adopted similar approaches that leverage unpredictability. It leads to more of a perceived reliance on the device. Which is why these devices have now introduced monthly subscription fees of anywhere from $5 to $30.

DATING

Dating apps are an exquisite case of using the loop to drive behavior and profits. Take Tinder. It's one of the highest-grossing mobile apps of all time. It made swiping left or right famous and revolutionized not just dating apps but apps in all domains. And it did this by taking the opportunity to fulfill one of our strongest drives—sex and companionship—and thrusting it into a scarcity loop.

Brian Norgard, the inventor of Tinder, said on a podcast, "When you think about Tinder, it's a[n] [unpredictable] reward game. Swipe-swipe-swipe, match. Oh my God, that was amazing. Swipe-swipe-swipe, hopefully I get another match," he said. "We took a lot of things from the [gambling] industry. . . . We did it in a subtle way but in a way where a lot of people saw paywalls."

Tinder is free, but these paywalls give users the opportunity to buy features that increase their chances of a match. Some slot machines have a similar feature: you can pay more per spin to boost the likelihood of scoring a bonus feature.

Then, once a person pays for the upgraded features, "you get more right swipes, more people like you, and immediately you have this delightful experience," said Norgard. "And you look at [what changed], and you can say, 'Well, that was because I purchased [the upgrades]!' And so if [companies] can tie cause and effect together in a consumer tool, big things happen. Big things happen."

A team of Canadian researchers found that singles became more reliant on dating apps after the pandemic. "It's all dating apps now," a friend told me of how most young singles meet significant others.

VIDEO GAMES

"In mobile games, rewards are given unpredictably." That's according to Daria Kuss. She's a scientist studying the psychology of technology use. She told me, "The suspense this creates keeps the player engaged. Because they know that if they just keep gaming, at some point in time they're going to get a reward."

The video game world has stolen so many tricks from casinos that researchers now have a term for the phenomenon. They call it "the gambling turn." It's happening in mobile, video, and computer games.

In a somewhat strange study, scientists in Canada took a group of people and strapped all kinds of monitoring devices to them. Heart rate monitors, stress monitors, and so on. They then had the people play *Candy Crush* for half an hour. The researchers discovered that the participants' bodies reacted strongest to the near misses embedded in the game. This led them to repeat the next game quicker and play longer. "In conclusion," wrote the scientists, "near-misses appear to have similar psychological and physiological impacts on *Candy Crush* players as slot-machine near-misses have on gamblers."

GIG WORK

The *New York Times* reported that Uber, as well as many gig economy companies like it, is "engaged in an extraordinary behind-the-scenes experiment in behavioral science to manipulate [drivers] in the service of its corporate growth."

Uber, for example, uses unpredictable rewards and suspense-inducing cues to nudge workers into driving extended hours and where the company wants them to be. It's also leaning into the psychology of near misses. When a driver wants to stop, Uber will bait her with an opportunity, alerting her, "You're just [insert some sum of money] away from making [some round sum of money]. Are you sure you want to go offline?" For example, "You're just $21 from making $250."

As the *Times* put it, the takeaway is that "pulling psychological levers may eventually become the reigning approach to managing the American worker."

NEWS

Elements of the scarcity loop exist in our 24/7, all-encompassing news environment. Media scholars say that political news before 2016 was kind of like a boring old slot machine while political news afterward was more like one of Redd's slot machines. This obviously started with president Donald Trump, who received four times more coverage than Obama. But politicians overall started acting more brash and unpredictable—keeping us in suspense about what they'd do, say, or tweet next and how that would make us feel.

Many of us fixated on the news, waiting for the next breaking alert. Because of this, political news outlets like CNN and the New York Times had their most profitable years ever.

Just as the casinos all shifted to Redd's slot machines, news also changed. Media analysts at the *Columbia Journalism Review* believe that the media has become so hooked on the ratings and profits of the post-2016 era that it will create a feedback loop. Politicians will behave more unpredictably to get more media attention. The media, in turn, will run stories differently to grab public eyes. A recent study suggests the analysts are correct. It found that the most unpredictable and contentious politicians on both sides got the most engagement on news sites and social media.

The scarcity loop occurs in many other domains. Brian Norgard, the Tinder founder, said that using it has "become a pattern a lot of [companies] follow." It's a most powerful driver of behavior.

The average person today spends anywhere from eleven to thirteen hours using digital media. That's on our phones, TVs, computers, and more.

But those are just the newest places the scarcity loop lives. The places where technology experts are programming it to manipulate our behaviors. Remember that the scarcity loop is ancient. Researchers are finding the loop in some of our most influential institutions, like our modern food, medical, and education systems.

They believe the loop is central to all sorts of damaging human behaviors and conditions ranging from overspending and obesity to addiction and burnout. It's even a part of how we think of and process happiness. And if we only focus on technology, we're missing much of the equation.

This might all seem like doom and gloom. But Sahl reminded me that the scarcity loop "works" because it's so engaging. It's a powerful game. There's nothing better at grabbing our attention, holding it, and pushing quick repeat behavior.

And engaging things are also often fun things. Not all my descents into the loop were negative. Not even close. For example, I now thoroughly enjoy playing slot machines once every month or two and stop once my $40 is gone. It's entertainment. It's no different from paying $40 to go to the movies, and I eat a lot less heart-clogging buttered popcorn. Sometimes I even win and walk away with more than $40.

I've also made interesting new friends and had positive interactions over some social media platforms. Of course, not everything I see on social media increases my sanity. And, yeah, I often find myself scrolling mindlessly for too long. But it's not all bad.

And maybe the stupid "strain scores" on my activity tracker lead me to be a bit more active.

Despite the criticism new tech receives, it's delivered many social benefits. To take just one example, some data suggests there was a drop in the suicide rate among LGBTQ teens after the introduction of social media. Social media connected LGBTQ teens in small towns. They could form a support network that wasn't previously there. Tech also allowed many of us to work and students to learn through COVID lockdowns.

Many tech ethicists, public thinkers, and politicians are calling for more tech regulation. They believe technology should be redesigned to be less engaging.

But do we really want that? Should Instagram or YouTube design itself to be boring? Do we want to live in a world without the opportunity to blow $40 at a slot machine? Or bet on our favorite NFL team to make the game more exciting? How far do we apply that logic?

I hope not too far. Is it possible to discover what *enough* of the scarcity loop is—but not overdo it?

William James, born in 1842, is considered the father of American psychology. His ideas laid the foundation of modern psychological thinking. James captured something profound about this brief stint of consciousness we all have and call life. In the end, he said, our life is ultimately a collection of what we pay attention to.

The gambling industry takes a lot of heat. But Sahl rightly pointed out that gambling is highly regulated. "There are strict laws on odds and what features slot machines can have," he said. Many popular products and services can use the loop however they want. And they have trillions of data points and decades of research, all showing what captures our most precious resources again and again and again. Some are using the loop for ulterior motives.

For example, the behavior-tracking company Nielsen recently released a report about changes in viewing and advertising habits. It highlighted a variety of wild case studies, saying that corporations are paying streaming companies like Netflix to interlace their products deeply into the storylines of our most watched shows. And this isn't just a product placement here and there. Brands are becoming a fundamental part of narrative arcs. Nielsen pointed to one company that became fundamental to the storyline of a blockbuster Netflix show and received "more than 8 million impressions to viewers 21+, a key age demo" for the product, Nielsen wrote. Here we think we're getting heartwarming and compelling television programming. Instead, it's also a multi-episode advertisement.

Or there are mobile games. "*Candy Crush* and other games like it are very much like slot machines; they just operate on a different economic model," Sahl told me. "They're trying to hook you, then get you to the point where you're stuck on a level and waiting for the game to give you a combination of symbols that when executed within your skill allows you to clear the objectives and move on. But that might require a combination of symbols that has a one in five hundred probability of falling. So they're bringing you to the point where you want to see the next level and will pay them $5 or whatever it is to move on." Much like dating apps.

This hook-then-charge model, the Nielsen report predicted, will become a rule in the future. "The reach and influence . . . in this emerging area is a tantalizing opportunity for brands that should be too good to pass on!" the report stated. Tantalizing, indeed. Embed this seductively simple, captivating, pleasurable loop in a product and watch attention and money roll in.

In 1928, the propaganda genius and father of public relations Edward Bernays wrote, "In almost every act of our daily lives . . . we are dominated by the relatively small number of persons . . . who understand the mental processes and social patterns of the masses. . . . We are governed, our minds molded, our tastes formed, our ideas suggested, largely by [people] we have never heard of.... It is they who pull the wires which control the public mind."

William James and Edward Bernays worried about forces co-opting our attention for ulterior motives a century ago. And that was long before anyone codified the hold the scarcity loop has on us.

It's a deeply attractive, ancient game that takes our hand and walks us into more. And there's a reason for that.

Why We Crave More

The novelist Margaret Atwood once said that humans have a "talent for insatiability." The pioneering psychologist Abraham Maslow described us as the "perpetually wanting animal."

They're not the first to notice this. Humanity has long acknowledged that we're constantly craving and consuming more and warned about the downsides of our desires. Consider Christianity's teachings around lust, gluttony, greed, and envy. And Buddhism's recognition that cravings and attachments cause all suffering. Or stories like the Greek tale of Midas, the Hindu tale of Kirtimukha, and the Chinese tale of T'ao T'ieh. These religious teachings and ancient myths warn us of the same phenomenon: when we give in to our boundless appetite for more, we end up devouring ourselves.

We seem to believe our internal and external conditions will be perfect and that we'll be able to finally "arrive" and rest once we fulfill our next want. This is a delusion. Once we've met our desire of the moment— no matter how big or small—our brains seem to produce the next one. It's a sense that we're one move from where we want to be. But once we make the triumphant play, the board expands.

I've seen this in myself. I spent much of my twenties chasing some perfect destination I believed was hidden in the next drink. I never found it. The search nearly killed me.

Once I got sober at twenty-eight, I did indeed stop devouring myself. That is, until I looked for something new to hunger for. My craving self never left. Rather, it shifted its focus to wanting more of other things: money and status, or stimulation from quick purchases or the next meal out.

But why? This question led me to the work of Leidy Klotz.

Klotz has a bachelor's degree in civil engineering, a master's in construction engineering, and a PhD in architectural engineering. All from the top engineering schools in the country. He's now a scientist and professor of engineering at the University of Virginia, where he investigates big questions about design and how it can improve the world and our experience in it. He's won tens of millions of dollars in grants and consulted for the Departments of Energy and Homeland Security, the National Institutes of Health, and the World Bank.

But the thing about being an expert in something is that our expertise can sometimes drag us so deep down the rabbit hole of what we've been taught and told and conditioned to do and think that we can't see out of it. Even the best and brightest default to behaviors and thought patterns they may not realize are, at best, not optimal. Or, at worst, just plain dumb.

Klotz learned about his blind spot a few years ago. It happened when he was out-engineered by a person who was shaky on toilet use and believed wholeheartedly in the Easter bunny. His three-year-old son, Ezra.

Klotz was showing Ezra the ropes of engineering. The two were building a bridge out of Legos. They'd made the span—the part of the bridge that people walk or drive on—and each was constructing a pillar to elevate the span.

But when they connected everything, the bridge was wonky. Ezra's pillar was shorter than Klotz's, which put the span at an awkward angle.

The PhD engineer had the solution. He searched the Lego bucket for more blocks to elevate the shorter pillar. Once he'd found the right blocks, he looked up and saw that Ezra had done something remarkable. He'd removed blocks from the taller pillar.

"And this was clearly the better thing to do compared to adding more blocks," Klotz told me. The bridge was not only flush. It was also more stable because it wasn't as high off the ground. Ezra's solution also used fewer Legos, which gave the two more resources to build an entire Lego city around the bridge.

It was a clarifying moment for Klotz. A peek outside the rabbit hole. Subtracting blocks hadn't even crossed his mind.

"And so I wondered," Klotz told me. "Do we overlook subtraction as a way to change things?"

Klotz removed Legos from one of the pillars so the bridge was wonky again. He then started taking the off-kilter bridge and extra Legos with him everywhere he went. When engineering students dropped in to his open office hours, the bridge would be waiting on his desk. When he'd have impromptu meetings with his fellow engineering professors, he'd remove the bridge and Legos from his briefcase. "Fix it," he'd tell them all.

These people had hundreds of years of schooling in problem solving by efficient design between them. But they all did the very un-Ezra thing. They added more Legos. All of them, bested by a three-year-old.

"I was interested in understanding why it seemed so counterintuitive to subtract," Klotz said. So he set up a series of experiments.

Each experiment gave a group of people a different problem to solve. In one experiment, the participants had to—you guessed it—stabilize a Lego platform. In another, they had to improve the flow of a miniature golf hole. Other experiments had the participants improve essays, recipes, or sightseeing itineraries. There were eight experiments in total.

In each, participants could solve the problem by adding or subtracting elements. The catch was that subtracting was always the most efficient solution.

Take stabilizing the Lego platform. It was an off-kilter one-legged table. Removing the single support would allow the platform to sit flush and sturdy. And the mini golf hole? Cluttered. It was like a hoarder hole. It was L-shaped, with angles and sand traps and the like.

Klotz even did what he could to dissuade addition and encourage subtraction. He "charged" the participants with fake money for each Lego they added to the structure or feature they installed on the mini golf hole. In some of the experiments, he even told people, "Keep in mind you can add things as well as take them away."

It didn't matter. The participants went to task—and immediately started adding. They tacked all kinds of extra Lego pillars to the structure. They installed windmills and angled bumpers and sand traps to the hole.

By adding more Legos and obstacles, the participants did solve the problem. But they did so inefficiently and at a higher cost. Their solutions used more time and more resources.

Klotz told me the immediate criticism of these experiments, especially the Lego one, "is that people say, 'Oh, well, we're conditioned to add and build with Legos.' And my snarky response is, 'Well, why are we conditioned to add and build with Legos?' But I get it. So that's why we did this grid study."

In "this grid study," Klotz asked participants to work with a ten-by-ten grid on a computer screen. Some of the grid's squares were randomly colored green, and the rest were white (picture a crossword puzzle). The goal was to put the green squares into a symmetrical pattern. Participants could do this by clicking on white squares to turn them green or on green squares to turn them white. Unlike building with Legos, Klotz said, this experiment was devoid of social context.

"You could solve the grid by adding twelve colored squares or by subtracting four," Klotz told me. "And we even told people, 'Solve this with as few moves as quick as possible.' Adding was obviously wrong." And yet the vast majority of participants clicked away. They added more green squares instead of turning the excess green squares white.

Klotz even flipped the equation in one of the experiments. He gave participants a grossly overpacked schedule for a sightseeing day trip to Washington, D.C. It contained twelve different excursions. He then asked the group, "How would you make this schedule *worse*?" The majority of participants found worse by removing excursions. By doing less. Even if, objectively, having fewer excursions would have freed the schedule and led to a realistic trip.

The takeaway from all of these experiments is this: In the human brain less equals bad, worse, unproductive. More equals good, better, productive. Our scarcity brain defaults to more and rarely considers less. And when we do consider less, we often think it sucks.

"People systematically overlook subtraction," Klotz told me. "If people were thinking about either addition or subtraction, then choosing to add, it would be one thing. But if people aren't even thinking of this basic option of subtraction, then that's a big problem. This is arguably the most fundamental question about how we change and make things better. Am I going to add, do more, or am I going to take away, do less? And we are finding that people systematically overlook the option of subtraction and doing less."

His research landed on the cover of the prestigious journal *Nature* in 2021.

Once we see this phenomenon, we see it everywhere. To toss out a few random recent examples: Federal regulations are seventeen times longer than in 1950. American homes are three times larger than they were in 1970. We own 233 percent more clothes than we did in 1930. Restaurant portion sizes are four times larger compared with 1950. Everything from our cars, to refrigerators, to microwaves, to coffeemakers is larger and packed with techy smart features (like, why does my dishwasher need to connect to the cloud?). Incoming university presidents are nearly ten times more likely to add new programs than they are to subtract programs that aren't working. "Yes, and . . ." is the rule of improv comedy, and the phrase has been adopted as a line by the business community. Entire industries like academia and medicine have experienced a 44 percent ballooning of professional administrators since the early 2000s. We create and consume ninety times more data and information today than we did just fifteen years ago. High-level employees spend an average of 130 percent more time in meetings today than they did in the 1960s. I'll stop now.

But there's no solid evidence that any of those tips into more are better. Consider meetings. More than two-thirds of managers said that most of their meetings are unproductive and inefficient. They said this new influx of meetings keeps them and their employees from completing important tasks, interrupts their thinking, and (quite counterintuitively) actually pushes their teams further apart.

No wonder we're experiencing what researchers call "time scarcity." It's a feeling that we don't have enough time. The truth is that we have more time than ever, thanks to advances in human longevity and the changing nature of work. Still, we cram our lives with so much compulsive activity, things "to do," that we feel pressed.

Our entire economic system, in fact, favors adding over subtracting. We judge a nation's power and prosperity by measuring gross domestic product (GDP). It's a measurement of all the stuff and services a country produces. The only way to improve the GDP? Add. Do more, make more, and extract more.

This adding phenomenon isn't new. What *is* new is that we now have many, many more ways we can add and pieces we can add with. Klotz explained that there is indeed an "array of biological, cultural, historical, and economic forces that push us toward more."

Natural selection is like the $3.99 steaks the old casinos once used to lure in gamblers. It's been grilled to death. It's probably the most scrutinized scientific theory of all time. It proposes this: Traits that cause us to have more offspring and survive become more prevalent over time. Traits that don't are weeded out.

Enter scarcity. Harvard anthropologists write that "natural selection acts most strongly not during times of plenty, but during times of stress and scarcity." They say scarcity fundamentally altered our minds and bodies. It built us to acquire and consume.

For example, humans began diverging from our ape ancestors sometime between 9.3 and 6.5 million years ago when a global cooling period led to a scarcity of food in the jungle. The apes who stood a bit taller and were better at covering ground were able to get more food, survive, and spread their genes. Over the generations, these apes began standing upright and walking on two feet because doing so allowed them to cover even more ground to avoid the dangers of scarcity. They became the first humans. We evolved to add, and adding nearly always made sense in our ancient worlds of scarcity, so it became our default.

But now decades of research have found that many of our biggest problems—at both the personal and the societal levels—come from our modern ability to easily fulfill our ancient desire for more. Scientists call this an evolutionary mismatch. It happens when behaviors and traits that help us in one environment hurt us in another.

Remember we now have an abundance of all the things we're built to crave. Everything from the food we eat, information we digest, social ladders we climb, and items we possess has shifted in ways that often clash with our evolutionary drives. Consistently obeying our drive for more in our world of more seems to be making many of us sick and miserable in ways obvious and unexplored.

Our craving for stimulation, calorie-dense food, ample possessions, information, status, and much more is backfiring in our world of concentrated drugs, junk food, online shopping, Google search, social media, and more. And corporations have created an entire arsenal of new technologies using the ancient scarcity loop to push us even further.

Unfortunately, we can't fix all these modern problems with some newfangled diet or workout routine, meditation commitment, media detox, or weekend of Marie Kondo–ing. Our brains, remember, are designed to constantly scan for and prioritize "scarcity cues." Those tip-offs in our environment that make us feel as if we don't have enough. They fire on the scarcity mindset.

"This scarcity mindset effect occurs even in relatively affluent, comfortable people," Kelly Goldsmith told me. She's a scientist at Vanderbilt University who studies how scarcity cues affect us. Her work shows that even the mildest scarcity cue—like running out of milk or ink in a pen—compelled people who had plenty to reach for more of all kinds of things and make decisions that hurt them in the long run.

"We found that when you threaten people's access to everyday things, they grab more for themselves and are less likely to give to others," Goldsmith said. More makes us feel safe. As if we were doing something to solve this perceived problem of scarcity.

And if we can't immediately slake our thirst for more, the thought of what we're lacking consumes us. The American Psychological Association explained, "Our minds are less efficient when they feel they lack something—whether it is money, time, calories, or even companionship."

Recall what the father of American psychology, William James, said about how our life is a culmination of what we pay attention to. Brainpower we could have used to plan ahead and solve real problems or just be satisfied and enjoy our present condition gets sucked into a vortex of craving. "This deprivation," wrote the scientists, "can lead to a life absorbed by preoccupations that impose ongoing cognitive deficits and reinforce self-defeating actions." That's scientist-speak for "we obsess over and do dumb stuff and that hurts us."

And yet we're not designed to know when we've tipped into excess.

Progress has provided us with more and better of everything. But we rarely stop and appreciate it. We're designed to habituate and move the goalpost. We want *even* more, *even* better everything.

As the MacArthur Genius grant winner and neuroendocrinologist Robert Sapolsky put it, "If we were designed by engineers, as we consumed more, we'd desire less. But our frequent human tragedy is that the more we consume, the hungrier we get. More and faster and stronger. What was an unexpected pleasure yesterday is what we feel entitled to today, and what won't be enough tomorrow."

The legendary basketball coach Pat Riley called this our "disease of more." Across his career, where he racked up eight NBA championship titles as a coach or GM, he noticed that championship teams across sports usually fail to win again the next year. "Success," he wrote, "is often the first step toward disaster." At first, pro athletes just want more wins. But once they win a championship, "more" shifts. They begin to focus their attention on a newly perceived scarcity. They now want more sponsorships, more playing time, more money, more individual recognition.

We can even watch this unfold in brain scans. Research from Cambridge University shows that our brain changes its chemical response to a reward based on our past experiences and expectations. For example, let's say we win $1 million. Sounds great, right? It is. But only if we anticipated winning *less* than $1 million. If we expected to win $1 million, then it's decent. But if we planned on winning $2 million, then winning $1 million is experienced as a disappointment.

Subtracting, of course, isn't inherently better. And neither is adding. It's just that we've been adding for a long, long time. And now more industries are discovering the scarcity loop and leveraging it to rapidly accelerate us further into more.

By defaulting to adding, we often make choices that are, at best, not optimal. Or, at worst, just plain dumb.

The answer, however, isn't always to subtract. Less can lead to its own set of problems. We need to ask the deeper question and consider how we can find enough. Not too much, not too little.

To find enough, I needed to uncover and learn more about all the places the scarcity loop exists, in technology and beyond. I needed to figure out where, why, and how far we're being pushed into more. I needed to get intentional with it. And understand, deeply, why we slip into the loop and the motives of the larger forces who are using it.

Thomas Zentall discovered something funny about his gambling pigeons.

"We typically keep them in small cages, and they seem to adapt fairly well to living alone in cages," he said. The numbers and data and figures all suggest their lives are good. "But we sometimes put them in large cages. They live more like they would in the wild. They can socialize. But not just socialize: The cage is designed to be more like their wild environments. They fly around a bit and go onto ledges, which is where they usually hang out in the wild."

He wondered what would happen if he presented pigeons with the two games after they'd spent time living like a pigeon in its wild habitat.

"After that, the birds actually start choosing optimally," said Zentall. They choose game one. The non-gambling game.

Zentall told me, "There's a model called the optimal stimulation model. It says that animals and we humans have a level of stimulation that we prefer. And when it gets below that, we search for stimulation. We found that if pigeons have had alternative forms of stimulation that were more like their life in the wild, it seems to reduce the likelihood that they will choose the gambling game for a substantial amount of time."

The pigeons could, in a way, realize that they had enough. They were fulfilled by what they had, craved less, and didn't want to escape into the scarcity loop.

"And when you think of humans today," Zentall continued, "I think a lot of us get bored with how easy it is to get resources. We spend less of our time exploring and foraging for food. We spend less time outside. Our social worlds have changed. So we search for other ways to fill this gap in stimulation, to distract or comfort ourselves." We're more likely to, like the unstimulated pigeons in small and sterile cages, fall into a manufactured scarcity loop. To fill our lives with mindless and counterproductive consumption. "When our needs aren't met," said Zentall, "we gamble, we shop online, we eat just to eat, we overuse social media, or even do drugs."

At the extreme end of scarcity brain, said Zentall, lies addiction. He's seen addictive behaviors in his degenerate pigeons that live in sterile cages, unstimulated. His colleagues have seen it in their lab rats and other animals that live the same. "And humans aren't much different, in many ways, from my pigeons," Zentall told me.

Zentall is alluding to a new and controversial theory of addiction. Since the 1990s, we've thought addiction is caused by unseen chemical phenomena deep in the brain. But more thinkers like Zentall realize that we have more in common with unstimulated pigeons and other animals than we might think. And their theories don't just apply to drugs and alcohol. It's a framework that can help us understand the root cause of any habit that delivers short-term comfort at the expense of long-term growth and fulfillment.

Now that I'd begun connecting the scarcity loop to scarcity brain, it was time to leave the safety of my home. If I really wanted to understand how to tame scarcity brain, I needed to meet people out on the edges. People finding answers in the real world and not just in sterile labs and over Zoom.

That extreme edge of scarcity brain seemed like a good place to start. First stop Baghdad.

Escape

Dr. Emad Abdul-Razaq was sitting behind an ornately carved walnut desk. Atop it was a plaque with his name and title inscribed in tightly swooping Arabic lettering. A three-by-five-foot Iraqi flag hung on a flagpole in a corner behind him. The wall to his left held bookshelves stuffed with psychiatric texts written in English and Arabic. Photos of the doctor with groups of powerful-looking government types in suits or military garb broke up the clusters of books.

The energy in the room was no good. The doctor was eyeing me as you might a person whose country recently invaded yours and is now asking for a favor. My fixer, Qutaiba Erbeed, was sitting in the chair next to me, smiling coyly and babbling excuses for me in Arabic. We shouldn't have been there.

Erbeed and I had spent the past few days madly driving around Baghdad. Erbeed, I quickly discovered, is a lunatic and hustler of the first order. He's in his late twenties, with a long mustache and pointy goatee like a pirate.

Before I arrived in Iraq, he'd sent me an itinerary for the week. It detailed all the places we'd go, people we'd meet, and precise times we'd meet them. The sheet also explained how we'd drive in a "very secure, top-of-line luxury SUV" and how I'd stay in "the most upscale and safe hotel in Baghdad."

Erbeed picked me up from the airport in a dilapidated, decade-old, base-model Hyundai. He then dropped me off at my hotel, a smoky and understaffed hole next to bombed-out buildings. Dust and water stains decorated the room. "Ah, yes, yes, we had a hotel problem," Erbeed said apologetically. "The nice hotel was full."

He then admitted that all of our "confirmed meetings" were just suggestions of what we might do in a perfect scenario.

When he sensed my hesitation about the car, the hotel, and the bluff of a plan, he took my hand. His oversized gold pinkie rings were cold to the touch. "It will be fine," he said. "Yes, *fine.*" Then he took me out to lunch. There he grossly over-ordered and screamed "Haram" at me, accusing me of breaking Islamic law if I didn't finish all the food and stuff myself.

The first few days, we mostly drove around Baghdad in Erbeed's neither secure nor top-of-the-line Hyundai as he frantically called government, military, narcotics, and health officials and attempted to turn all his "proposals" into realities. And so I thought this trip was a wash.

But just as Erbeed's grift worked to get me to Iraq, it eventually began working on others. The man was talking, always talking. He talked us into situations we wanted to be in and out of those we didn't. He talked us into terrorist and drug kingpin holding cells, narcotics enforcement outposts, addiction rehabilitation centers, off-the-books meetings with army intelligence officers, and other places on the front lines of Iraq's emerging battle against a curious new drug. All the while, he also talked us *out* of some tricky situations like confrontations at security checkpoints and with jihadi militia members who'd prefer to see the United States and everyone in it wiped off the map. We saw many guns.

On the third day, Erbeed managed to secure meetings with Iraq's Ministry of Interior, which oversees policing and border control. But then an epic sandstorm engulfed the city. Visibility was down to twenty yards. Iraq treats sandstorms the way other countries do blizzards or hurricanes. Schools, government buildings, and nonessential businesses all closed. Things were looking hopeless.

But the following morning, he talked us into that police compound. As he jabbered, facing plenty of resistance, I realized I needed to take things into my own hands. I was able to locate some in-country journalists and contact them via, of all things, Instagram. One passed along Dr. Emad Abdul-Razaq's phone number. After our time at the police compound, Erbeed called. The doctor picked up to tell him to text instead. As Erbeed was texting the doctor, he began to grin.

"He thinks I am another man with the same name," said Erbeed. "But he agreed to meet with us." When I balked, Erbeed shut me down. "No, noooo, the doctor will talk. It will be fine. *Fiiiiiine.* He will talk. Yes, he will talk. You must worry about other things, like what we will have for lunch." I'm not proud to say I was desperate enough to agree.

We then careened through the heat of Baghdad. Past manned sniper towers on Damascus Street and militia members lined up and protecting Tahrir Square on the banks of the Tigris River, a hot spot for protests and attacks.

As he drove, Erbeed was juggling a cell phone in each hand. He'd talk on one phone and text on the other as he clumsily steered the vehicle with his knees. The man might be the worst driver in the Middle East. Just that morning, he tore open the sidewall of one of his tires by trying to jam the Hyundai between a curb and another car.

He looked at me as air whooshed from the tire. "You are American. This means you like the cars. You know how to change a tire?" No AAA here. So I changed the tire in the hundred-degree heat. As I twisted off lug nuts, all seven million residents of Baghdad seemed to drive past and curse at me to get out of the road.

Soon after, Erbeed sideswiped a taxicab that stopped in front of us while he was texting. Not twenty minutes later, he converged with an SUV while blabbering, bouncing off it like real-life bumper cars.

In both accidents, Erbeed slowed our vehicle, rolled down my window, then leaned over me to scream at the other driver in Arabic. Then he floored it. Neither car stopped to exchange insurance information or anything like that.

"What did you say to them?" I asked.

"I say, 'Why you in my way?!?!?'" he replied.

And then there I was. Sitting in that con of a meeting with the head of psychiatry of all of Iraq, whom Erbeed had topped off with some bullshit or another. I had to hand over everything I had with me to heavily armed guards at the office's entrance. Passport, visa, computer, burner phone, GPS locator, various tools I could use to possibly escape if I was kidnapped, and $5,000 cash. I now had a wallet, notebook, and pen.

Dr. Abdul-Razaq took his dark gaze off Erbeed and shifted it over to me. "He says you are a professor," Dr. Abdul-Razaq said in English, motioning to Erbeed, who was grinning like a door-to-door salesman. "Do you have identification?"

"Ummmm," I mumbled as I removed my wallet. The request felt equal parts theatrical and demeaning.

I handed the doctor my worn ID card from the University of Nevada, Las Vegas. He tipped his glasses and scrutinized its information as if he'd be tested on it later. "It says your title is 'Faculty,' not 'Professor,'" he said and began to shoo us out of the room. I explained that in the United States, "Faculty" indicates all university teaching and research positions.

"What academic department are you in?"

"Journalism," I said.

He rolled his eyes. "If you want to know about drugs and addiction in Iraq," he said, "download the 2014 Iraqi household survey on drug use. It is on our website." He motioned us to the door again.

But the thing about that interaction compared with most others I had in Iraq is that there wasn't a gun in sight. No AK-47s. No Glocks. They were all down the hall. I'd traveled seventy-five hundred miles to get there so I could understand the extremes of scarcity brain. And this unarmed doctor-slash-government official could help me do just that. So why not press him?

I told the doctor that data from 2014 is too old for me to rely on. Too old for him to rely on, too. A lot has happened in Iraq since then—the rise of ISIS, a U.S. military exit. And I was interested in learning more about a new drug called Captagon that I'd heard was overtaking Iraq.

Just as the doctor started to look at me the way the jihadis did, I began to admit that my own country had a drug problem—worse than Iraq's. And I believed the new frontier of drugs in Iraq could help me understand addiction, an issue that cares not about borders and backgrounds and bank accounts or anything like that. And that might shed light on all sorts of other universal problems.

Maybe it was my use of sciencey terms like "data." Or my admission that my own country was in some ways as messed up as his and also had a massive hand in messing his up. Whatever it was, the doctor's look became less hostile. He dropped my faculty-not-professor identification card on his desk. Then he leaned back into his leather chair and stroked his Tom Selleck–like mustache.

"Yes, there's been a vast increase in drug use in this country," he said. "It began to increase after the American invasion in 2003. This invasion destabilized the country and opened the door for the circumstances in which people would take drugs. Then, since the war with ISIS in 2014, we've had a massive increase in the supply of drugs. Especially stimulants. Like Captagon. The Captagon problem is bad and growing."

In 1961, a German pharmaceutical company invented the drug Captagon. It marketed Captagon as an alternative to amphetamines like Adderall, Dexedrine, and Benzedrine. Captagon increased focus and induced euphoria, so it was used to treat children with ADHD and adults with depression.

In the 1970s, doctors began noticing that Captagon worked too well. Its tendency for abuse and addiction outweighed its medical benefits. Using the drug recreationally became popular in oil-rich Persian Gulf countries like Saudi Arabia. Islamic countries generally follow strict rules around haram, or forbidden acts. Taking drugs and drinking alcohol is haram. But because Captagon was a legal and medically backed prescription pill, some saw using it as a way to get high without breaking haram. This isn't unique. Utah, for example, has one of the highest prescription drug abuse rates because Mormons aren't allowed to drink alcohol or coffee or use nicotine.

By the 1980s, the United Nations had listed Captagon as a scheduled substance. Countries worldwide began prohibiting its production and use. But the Middle East couldn't kick its habit.

To supply the demand, the Syrian military and the Lebanese terror organization Hezbollah stepped in. In the 1990s, they began setting up small clandestine labs along the Syrian-Lebanese border to fund military campaigns. But after the ban, the supply of Captagon's original active ingredient (fenethylline) had run out. So they filled the pills with random mixes of methamphetamine, amphetamine, caffeine, heavy metals, and any other cheap stimulant they could find.

The 2011 Arab Spring movement set Syria into a civil war that continues today. It pitted President Bashar al-Assad's Shia Muslim regime against groups like al-Qaeda. In 2014, ISIS rose in Iraq and entered the conflict. They fought both the Iraqi government and the Assad army in an attempt to create a hard-core Islamic State in Iraq and Syria (hence the acronym ISIS).

Before I arrived in Iraq, I spoke with Caroline Rose. She researches the Captagon trade at the New Lines Institute, a geopolitics think tank in Washington, D.C. She told me that to fund war efforts, the Assad regime took over pharmaceutical plants and set up mega-labs. "They employ a large number of people to produce Captagon in industrial-sized amounts," she said.

Today, there are at least fifteen major Captagon production facilities in Syria that the government supports. "In Syria, Captagon production is mostly affiliated with the Fourth Division," said Rose. They're an elite military unit within the Syrian Army—akin to SEAL Team Six.

"And then you also have organizations with deep ties to Iran's Revolutionary Guard Corps who help," said Rose. Iraqi state-sponsored militias like the Popular Mobilization Forces, which receives training and funding from Iran, also aid in the Captagon trade.

The day after our meeting with Dr. Abdul-Razaq, Erbeed and I traveled to the Al Mansour mall in Baghdad. Picture a standard, multilevel shopping mall with a big food court complete with a Burger King and knockoffs of the American fast-food restaurants that wouldn't franchise to Iraq.

We met with Ehab and Nader (whose full names are withheld for security purposes) at a hookah lounge on the mall's top floor. They're both intelligence officers in the Iraqi Army who run antidrug operations against ISIS and other smugglers along the Syrian border. We surrounded a table overlooking the sprawling lights of Baghdad. The men were smoking lemon shisha tobacco from a traditional hookah while I sipped a thick cardamom coffee.

"We started noticing more Captagon beginning in 2014, during the war with ISIS," Ehab explained. "ISIS fighters would take it to kill fear, to feel stronger, and stay up longer. They took it to be ready to die. Many suicide bombers would take Captagon."

Because of the drug's connection to ISIS, Western media began calling Captagon the "Jihadi pill." But Captagon quickly became popular among fighters on both sides of the war. "Many Iraqi infantry soldiers are now using Captagon or other amphetamines," said Nader.

Amphetamine use among soldiers isn't new. A paper in the *Journal of Interdisciplinary History* reports that both Allied and Axis troops in World War II used different amphetamines because the drugs, the scholars wrote, "increased confidence and aggression, and elevated 'morale.'"

Nader and Ehab are both frequently deployed to the border, fighting a war against a surging drug trade. They echoed Rose's point about the role of Iraqi militias in the trade. "A lot of militias are crooked. They take bribes or ignore the problem because some of the money from drugs flows into Iran," Nader said.

It's an entire military-narco-industrial complex of jihadi actors who produce and smuggle Captagon, which has become the most in-demand drug in the Middle East. The demand is a specter that's spreading and swelling into new regions. The value of Syria's illegal Captagon industry is now forty-five times that of its entire legal export industry. But unlike the cartels of Mexico—who make and distribute drugs for wealth— riches from Captagon are used for terror campaigns. To fund organizations implicated in war crimes and terrorism.

And the numbers are staggering. Authorities in Malaysia and Italy recently discovered two shipments of 100 million pills, each being routed through their ports back to the Middle East. Experts estimate between just 1 and 5 percent of drugs that cross borders are confiscated. At this very moment, billions of Captagon pills are circulating throughout the Middle East. Billions.

Rose recently analyzed all the numbers. In 2020 the trade was worth $3.46 billion. In 2021, $5.7 billion. In 2022, it reached more than $10 billion. In 2023 and beyond, it'll likely rise well beyond that. But some analysts believe those figures are too conservative. The market may be three times bigger than that.

The United Nations Office on Drugs and Crime reports that drugs used to pass through Iraq on their way to other Middle Eastern countries. But drugs began pouring in and staying a decade ago. In just a few months in 2021, Iraqi forces seized twenty times the amount of Captagon they'd captured in 2019 and 2020 *combined*. Recent seizures, however, were overwhelming all of that, said Ehab.

"The border with Syria is very long," explained Nader, blowing lemony smoke from his mouth. Smugglers are getting creative. "We recently stopped a shepherd crossing the border from Syria into Iraq. We found that he'd cut open the stomachs of his sheep and placed packs of Captagon inside the sheep stomachs, then sewed them back together."

Violence around the trade isn't only in Baghdad, Ehab told me. "We've had many of our soldiers killed," he said. Traffickers are now taking a kill-first, evade-second approach to smuggling Captagon across borders. Dealers tote AK-47s, rocket-propelled grenades, sniper rifles, and improvised explosives. The Iraqi colonel Ziyad Al Qaisi recently said that although terrorism and drugs are two sides of the same coin, combating the drug trade is becoming more dangerous.

When I arrived in Iraq, a wave of Captagon was sweeping over Baghdad. Days earlier, authorities had stopped a transport truck stuffed with 8 million pills on its way to Baghdad from Syria. They also broke up a local drug ring planning to distribute 6.2 million Captagon pills.

A woman in a black hijab entered Dr. Abdul-Razaq's office. She placed small paper cups of copper-colored Ceylon tea flavored with cardamom on a coffee table in front of Erbeed and me. "As-salamu alaykum," said Erbeed. "Peace be upon you."

The doctor nodded as the woman turned and exited the room. Then he began. "In Iraq, we have many victims of trauma. This trauma started in the 1980s, with the war with Iran," he said. In that war, Saddam

Hussein invaded Iran, which had just experienced an Islamic revolution. Because Iran was in a state of postrevolutionary chaos, Hussein thought he'd roll through the country and score a quick and easy takeover. The war lasted eight years, ending in a stalemate.

"One million people died in that war," said Dr. Abdul-Razaq. Two years later, in 1990, Hussein invaded Kuwait, leading the United States and other countries to intervene and begin the Gulf War. Then there were uprisings against Hussein, like the Iraqi-Kurdish Civil War from 1994 to 1997, which led to genocide. Hussein is said to have committed genocide on a quarter million Iraqis during his reign.

In 2003, the United States and its allies invaded Iraq. The ensuing war ousted Hussein and lasted, officially, until 2011, although the ramp-down was slow. Some estimates suggest a million people died in that war. ISIS exploited rising animosity toward the West and rose from the rubble.

From 2013 to 2017, the heinously violent organization led a reign of terror in an attempt to create a caliphate. The Battle of Mosul, for example, is considered the fiercest urban combat since World War II. Still today, Iraq fights an ISIS insurgency, with bombings and violent assaults and executions happening every other day and growing. The economy is also in ruins. "Foreign investment is minimal, and unemployment is high," said Dr. Abdul-Razaq.

"So I think that addiction is most often a function of circumstances," said the doctor. "If the conditions are right and drugs are available, drug use rises. People use drugs for good reasons. Drugs are an easy way to escape, feel empowered, cope with life, and survive. Some hardworking Iraqis use drugs to stay awake and work longer hours."

The doctor was alluding to the same theory of addiction Zentall did. We've long viewed substance use as a bizarre anomaly—a function of either broken morals or broken brain chemistry. But there is a growing recognition that drug use in our past was nearly always good. It helped us survive.

Our modern tip into abundance then shifted the nature of substance use. And yet the brain of an addicted person still, in a way, views their drug use as a survival mechanism.

If we can unpack how and why that occurs, we can understand the internal and external conditions that lead people to use Captagon in Iraq and beyond. And this can help us understand the root of many of the habits we most want to change. It can help us understand a key reason why, when, and how we fall into the scarcity loop. And how we can find enough.

If you look up the definition of "addiction" in the *DSM-5,* which is the bible psychiatrists use to treat patients, it isn't there. The *DSM-5* doesn't use the word "addiction." And that is, they write, "because of its uncertain definition." People who have dedicated their lives to studying addiction gave me all sorts of different definitions. But the scientists and practitioners I spoke with all generally circled the same idea. Addiction is chronically seeking a reward despite negative consequences.

Under that framework, our ancestors were all addicts. Early humans went on epic hunts despite incredible dangers. They explored new territory, searching for more, despite not knowing what harms lay in the unknown. They braved treacherous weather, wildlife, and landscapes, hoping for a reward in the form of food and safety. If our species wasn't willing to chronically persist in the face of negative consequences—say, breaking an arm on a hunt, getting lost and hungry exploring for food, and then repeating it the next day—we would have all died off.

Brain scans even show that addiction lives in the same system of the brain involved in love, whether with a partner or our children. "That system quite naturally evolved to create compulsive behavior despite consequences," Maia Szalavitz, an addiction researcher, journalist, and author of *Unbroken Brain,* told me. "You'd never maintain a relationship with a partner or deal with all the diapers and crying and frustrations that come with raising a kid if there wasn't some deeper reward that allowed us to persist despite negative consequences."

In fact, the "addicts" of our tribe—the people who persisted the longest despite the most heinous conditions and consequences—were probably the most successful. Until, of course, it killed them.

Enter drug use. Drug use isn't new, and it isn't unique to humans. Scientists now know that the use of psychoactive substances has "exceedingly deep evolutionary roots." It goes back to multicelled organisms that first appeared 500 million years ago.

I spoke with Moira van Staaden, an evolutionary biologist at Bowling Green State. She told me that drug compounds like amphetamines, cocaine, opioids, nicotine, and alcohol first developed in plants as defense mechanisms against insects.

It worked like this: A bug would begin to eat the plant. But soon that drug compound within the plant would affect the bug's behavior in a way that protected the plant. Van Staaden explained, "For example, picture an insect that is camouflaged. If that bug eats a plant that contains amphetamines and those compounds cause the bug to suddenly start moving faster, that bug is going to break its camouflage and be seen and eaten by predators much faster and easier."

Other compounds, like opioids, might slow the bug down. Other chemicals might affect the bugs' drive to procreate, leaving just one inebriated bug to eat the plant and not any new generations of them. The high from these drug compounds also had to be rewarding. If the bug didn't die during its first encounter with the substance, it had to want to come back for another round.

Van Staaden's experiments show that all of the compounds we humans use to change our mental state—cocaine, opioids, amphetamines, alcohol, nicotine, and more—have the same basic effects in bugs as humans. For example, when she doses crayfish with amphetamines, they become hyperactive. When she gives fruit flies alcohol, their flying pattern becomes the aerial equivalent of stumbling. When she pumps crayfish with opioids, they slow down and lose interest in procreating.

"So drug compounds did not evolve to be directed at us humans," van Staaden said. "They were directed at insects." But the genes that first appeared in insects spread over hundreds of millions of years and were passed down to us.

"Certainly, at a structural level, we are very different" from the bugs that first appeared 665 million years ago and still exist today, like crayfish, van Staaden told me. "But if you look at receptors and biochemistry—at a chemical and molecular level—we share a good proportion of our genes with insects, 80 percent on average at all levels. And that is why, therefore, we are also vulnerable to the same psychoactive compounds within the plants. Humans are, in essence, collateral damage of this."

Our species, *Homo sapiens*, has been taking mind-altering substances ever since we evolved from our common ancestors. And those ancestors also took mind-altering substances. As did their ancestors. And so on down the line to those ancient bugs.

Yet we aren't bugs. We're much larger and smarter. We learned to use those plant compounds as tools. Scientists at Stanford say the substances and behaviors we consider most addictive today helped our early ancestors survive the crucible of evolution.

Consider alcohol. Early humans valued fruit over most other foods because fruit is high in sugar and calories. As wild fruit falls from trees and ripens, it mixes with yeast that occurs naturally in the air and begins to ferment. Scientists at UC Berkeley say this fermenting fruit emitted a funky smell that helped early humans find it. And it also created low levels of alcohol within the fruit.

Eating this boozy fruit triggered what scientists call the "aperitif effect." Studies show that people eat anywhere from 10 to 30 percent more food after drinking alcohol. So it compelled our ancestors to find and gorge more of the fruit to bulk up for leaner times. The buzz further rewarded us. Alcohol even kills germs. So this boozy food was less likely to contain bacteria that might make us sick.

This means that in the past, alcohol equaled survival. But the alcohol levels in this wild fermenting fruit were so low that we'd fill up long before we'd start slurring our words.

The same rule follows for stimulants, opioids, tobacco, psychedelics, marijuana, and more. Cocaine comes from coca leaves. These leaves contain low levels of the chemical cocaine—just a fraction of a percent. Chewing the leaves killed our hunger and increased our stamina and focus during our long and hard hunts and searches for food. Other naturally occurring stimulants like khat, ephedra, and caffeine-containing plants did the same. We used opium from poppies during ceremonies and to blunt the pain of injuries we'd acquired during the hunt, like an ancient Advil. Tobacco sated our hunger and honed our focus on acquiring food when food was scarce. It also contains chemicals that ward off stomach parasites and infections. We used psychedelics ceremonially and medicinally, and possibly to help us dream up new perspectives on survival and ways of seeing the world. Marijuana treated a variety of medical ailments.

For nearly all of human existence, these substances were relatively weak and scarce. We could consume only so much of their active ingredient every so often. The highs were molehills rather than Himalayan mountains.

But, relatively recently, we took each of these plants' psychoactive components, concentrated them, and scaled up their availability. We created distilled spirits, concentrated powdered cocaine, methamphetamine, Captagon, heroin, modern cigarettes, and psychedelics and marijuana we can eat in candied form or inhale in vaporized liquid.

In the last few years, many illicit drugs shifted from being grown on the land—cocaine from coca plants, heroin from poppies—to being formulated in a laboratory.

The supply of lab-made drugs like meth, Captagon, and fentanyl is now larger than ever because these drugs are more than twenty times more profitable than land-grown drugs. They're also up to eighty times stronger than their land-grown equivalent. Massive supplies have driven prices down to "an all-time low," said researchers at John Jay College. Drugs like heroin and meth are more than ten times cheaper than they were in the 1980s.

Use and overdoses have since boomed. Meth overdoses in the United States are up nearly sixteen-fold since 2000. Fentanyl is the primary reason drug overdose deaths recently exceeded 100,000, up roughly sixfold since 1999.

Robert Sapolsky, the Stanford neuroscientist, summed up our modern landscape of psychoactive substances this way: "Once we had lives that, amid considerable privation, also offered numerous subtle, hard-won pleasures. And now we have drugs that cause spasms of pleasure and dopamine release a thousandfold higher than anything stimulated in our old drug-free world."

Although we have more drugs that are more powerful than ever, addiction isn't anything new. Thinkers going back to the ancient Greeks (and probably much earlier) have tried to figure out why some people become addicted and others don't. All these thinkers have come up with more than a hundred explanations for why. But the last century has left us with two dominant schools of thought.

The first views an addict as a bad person. It sees addiction as a selfish and destructive personal choice. The War on Drugs, for example, is based on this model. At its height in the 1980s, the War on Drugs tossed one million Americans into jail each year for offenses like possessing a small amount of marijuana, cocaine, or heroin.

But by the mid-1990s, we realized that we were losing this war. We'd spent trillions combating drugs worldwide and arresting and jailing people for possession. But drug use rates weren't budging. The approach was ruining lives and disproportionately hurting minorities. Research shows, for example, that most crack users in the 1990s were white, but 90 percent of people sentenced for crack possession were black.

So in 1995, a group of researchers from the National Institutes of Health's National Institute on Drug Abuse (NIDA) convened to search for a solution. Neuroimaging was a hot field at the time. The technology, which captures images of how our brains react to different situations, had recently improved drastically. New scans showed that the brains of chronic drug users responded differently to drugs than nonusers'.

Recall that dopamine isn't the pleasure chemical. Instead, one of its roles is to help us pursue pleasure by creating cravings.

Neuroscientists call the brain system that *does* produce the pleasure our "liking" system. Our liking system is composed of a series of tiny "hedonic hot spots" in our brain. When we do something enjoyable, these "liking" hot spots get hammered with chemicals such as mu opioids and endocannabinoids. These are naturally occurring brain chemical versions of heroin and marijuana. Unsurprisingly, this creates a fantastic feeling and causes us to "like" whatever we just did.

But when neuroscientists studied drug addicts' brains, they found something strange. The "liking" or pleasure hot spots in many drug addicts' brains often didn't "like," or activate, when they took their drug of choice. But their brain still released high levels of dopamine, leading them to continue craving the drug. The scans suggested that drug addicts craved the drugs but didn't like them once they took them.

With surprising findings like that in mind, the leaders of NIDA came up with the second model of addiction. It sees addiction as a disease of a broken brain. The idea is that addictive substances "hijack" our brains. NIDA argued that addicts are helpless passengers of this hijacked and runaway freight train that is their brain. Even if an addict chooses to stop, said NIDA's director, Nora Volkow, she won't be able to. There is no choice. This is because, as Volkow wrote in the scientific journal the *Lan-*

cet Psychiatry, "specific molecular and functional neuroplastic changes at the synaptic and circuitry level . . . are triggered by repeated drug exposure." This idea, called the brain disease model of addiction, has been at the front of scientific and policy circles ever since.

Dr. Abdul-Razaq and I had been speaking for half an hour. Erbeed, for once, wasn't having to translate. So he relaxed and pretended to care about the conversation while sipping his tea and sneaking looks at his phones. I asked the doctor about his thoughts on the brain disease model.

"Eh, it's interesting," he said. He doesn't dispute that an addict's brain responds differently to drugs than a nonaddict's. Of course it would, he said. But he worries that this complex neuroscience doesn't do much to help people.

The doctor is one of a growing number of people who work in the trenches helping addicts. They don't believe addicts are bad people. But their experience also doesn't suggest that addicts are helpless, hijacked passengers.

"Between the poles of diseased and depraved," wrote one of these new thinkers, "is an expansive middle ground of experience and wisdom that can help explain why millions use [drugs and alcohol] to excess."

Before traveling to Iraq, I spoke with many experts around the world who are exploring that middle ground. Dr. Sally Satel is an addiction psychiatrist and professor at Yale University. She told me she can see how the idea that addiction is a brain disease is helpful in getting addicts professional help rather than time in jail. "If I'm speaking to a judge, sheriff, or insurance adjuster, then I say, 'Yes, addiction is a disease,'" she said.

But none of the addicts she worked with seemed to be hopeless passengers of a hijacked brain. "Yes, addiction changes the brain," Satel told me. "But so does everything else. This conversation we're having right now is changing our brains. The crucial question is not whether brain changes take place. They do.

"The real question," said Satel, "is whether those brain changes obliterate the capacity to make decisions. The answer to that question is no. Choice might be constrained. But people *are* capable of breaking through the neurochemical storm and changing their behavior. It's possible to change. Everything is possible."

No science has ever been able to document specific changes in the brain that occur *because* of drug use and lead to addiction. And part of the reason we believe drugs hijack our behavior comes from experiments on lab rats that were bred for experimentation and raised in small cages, often alone. In these studies, the rats can hit a lever for a dose of cocaine. And they love it. These rats press and re-press the lever. They'll become so obsessed with getting high that they forget to eat and they die.

But what doesn't get talked about is what happens when you give these lab rats a safer and more comfortable, ratlike environment. The kind they'd have in the wild. A big park that has holes to hide, nests, wood chips, exercise wheels, plants, other rats, and more. They stop pressing the lever. Just as Zentall's pigeons stop gambling after living a wild pigeon life.

The addiction establishment often overlooks the reasons humans stop pressing the metaphorical lever. NIDA, for example, quietly acknowledges the environmental role of addiction. But the idea is pushed to the background. Most of its funding goes to researching the brain disease model. Most of its public information covers addiction neuroscience.

NIDA argues that addiction is a life sentence. It says "addiction is . . . a chronic and relapsing disorder," because its internal data shows that 40 to 60 percent of its addicted patients relapse. But people kick addiction for all kinds of reasons. And it happens far more frequently than official data might suggest.

"There's an idea called 'the clinician's illusion,'" Satel told me. It refers to the idea that addiction researchers and doctors tend to study and see only the most challenging cases, the people who can't stop despite losing everything. And this leads researchers and doctors to believe that all cases of addiction are as hopeless.

But research on regular people struggling with substance abuse suggests a much brighter outlook. Satel pointed me to one study that surveyed roughly twenty thousand people. It found that 75 percent who reported struggling with drugs before age twenty-four no longer used drugs by age thirty-seven. Another extensive survey found that over ten years 86 percent of people struggling with an addiction got clean.

Satel began firing off counterexamples from the people she'd worked with. People who stopped because they finally got tired of being homeless. Or because they got fired from a job or got a DUI. Or because their drinking caused them to miss their kid's soccer game. Or because they had a child or a new job opportunity. Or because, as the psychologist William James wrote, they'd had a "vital spiritual experience" that kicked them out of the cycle of addiction.

"And even if we accept that addiction is a brain disease," Satel told me, "we need to be real about what kind of disease it is. It isn't a brain disease like Alzheimer's is. There are no amount of incentives or punishments that could change the course of a disease like Alzheimer's. But there's plenty of evidence that people with addiction respond to incentives."

Take what happened to soldiers serving in Vietnam. Military doctors estimated that between 10 and 25 percent of American soldiers deployed in Vietnam were addicted to heroin. It was so bad that in May 1971, the *New York Times* ran a front-page story titled "G.I. Heroin Addiction Epidemic in Vietnam."

President Nixon didn't want to let these addicts back into the United States, so he launched Operation Golden Flow. The deal was simple. If the soldiers wanted to return home, they had to provide a clean urine test. If addiction obliterates choice and relapse is a foregone conclusion, then most of those addicted soldiers wouldn't have been able to overcome their broken brain and would have been left in Vietnam.

But nearly every soldier produced a golden flow, a clean urine test. And this wasn't a temporary stay. Once they were home, heroin lost its appeal for the majority of these GIs. Only 5 percent of the soldiers who were addicted while in Vietnam relapsed within a year of returning to the United States. The relapsers tended to be soldiers who had used drugs before the war.

The results shocked the scientist who headed the study. His findings, he wrote, "ran counter to the conventional wisdom that heroin is a drug which causes addicts to suffer intolerable craving that rapidly leads to re-addiction." Something about the incentive of leaving the hell of war shifted these vets' psychology. They were like Zentall's pigeons and the lab rats who moved out of the cage and into a world more like their natural environment.

This phenomenon also plays out in modern laboratories with smaller incentives. One study offered opioid addicts drugs or a voucher. If the addicts returned in a handful of days and produced drug-free urine like the soldiers in Vietnam, they could turn in the voucher for $25. About 75 percent of the study participants took option two and got paid.

In another study, researchers at Columbia University gathered a group of people addicted to crack cocaine and offered them two choices. They could smoke varying amounts of the drug or receive $5. The group overwhelmingly chose the money, although more people began picking crack once the hit got big enough. In other words, the researchers noted, the drug users made rational economic decisions.

Other research has shown that drug addicts' brains aren't as fundamentally broken as neuroscience scans might lead us to believe. The same Columbia researchers found that regular drug users made equally good decisions in a decision-making task as nonusers.

"The effect of incentives decays unless greater changes are made in their life," said Satel. "But the point is that addicts can respond to incentives. Their brain is capable of decision making. Yet if you take the logical extreme of the brain disease model that NIDA has put forth, that would never be able to happen. Really, I'm just against portraying people as helpless."

So far, there isn't a miracle cure like a pill or procedure for addiction. Recovery takes effort. But fully accepting the brain disease model can, for some, kill motivation to put in the necessary hard work.

For example, scientists at the University of New Mexico analyzed alcoholics in recovery for more than a year. The top reason for relapse was believing addiction is a disease. The relapsers said they didn't see the point in struggling against a disease without a medical cure. This viewpoint can also lead would-be lifelines to give up hope. Other research found that the more a drug user's family members believe addiction is an insurmountable disease, the more likely they are to distance themselves from the user.

But even brain scans suggest that a person in the grips of addiction *can* affect their brain chemistry. Researchers at Yale and Columbia had a group of smokers who were trying to quit watch a video of people smoking. Because of this cue, their brain pumped out dopamine that led them to crave. But when the smokers considered the long-term problems that come from smoking, like cancer, their craving circuitry cooled down and the area of their brain that controls long-term decisions activated. Similar findings have been shown with cocaine addicts. And this can work for any craving. If we consider our craving's downsides—weight gain from eating a second donut, anxiety from another social media binge—it can shift our brain chemicals and help. Which is to say you're not a slave to your brain chemicals.

"The implication of blaming addiction on brain chemicals like dopamine," Satel told me, "is that addicts should be taking dopamine blockers," medication that blocks the release of dopamine. "But that doesn't work."

Satel and Abdul-Razaq aren't alone in their ideas. Satel has worked with other psychiatrists, clinicians, nurses, and even mothers of addicts who are also against the idea that addicts are hopeless. There were Pam, Beckey, Lisa, and Sharon, whom she befriended in Ironton, Ohio, an opioid-addled town where Satel worked with patients for a year. These women "were genuinely interested in the question of addiction," Satel wrote about her experience there. "And they were fed up with the false choice routinely thrust upon them . . . : either endorse addicts as sick people in need of [medical] care or bad actors deserving of punishment."

In that expansive middle ground lies the original reason that humans and other animals began using psychoactive substances in the first place millions of years ago. "The brain disease model," Satel wrote in a study in *Frontiers in Psychiatry,* "obscures the dimension of choice in addiction, the capacity to respond to incentives, and also the essential fact people use drugs for reasons."

It's an idea Dr. Abdul-Razaq echoed. "People here use drugs for a good reason," he told me, leaning back into his chair and stroking his mustache. Every Iraqi he's worked with who has become addicted to Captagon, he said, began taking it because it initially improved their life. Many took it to self-medicate to deal with the traumas of war or enhance an experience. "But we also have lots of truck drivers or bakery workers who use it to stay awake and work long hours," said the doctor. A more extreme version of your morning coffee.

This, too, has been the American experience. Remember the formula: *We do a thing. If the thing was good, we're now more likely to do the thing again in similar conditions.*

And although it isn't talked about much, addiction falls deeply into a scarcity loop. As someone who is in recovery and talks to many others in recovery, I've experienced this. Drinking allowed me to let down my guard and behave in a way that, when sober, felt uncomfortable. My alcohol abuse fell into its own strange scarcity loop.

I longed not so much for alcohol as for the psychological state where I felt as if anything were possible. From wider experiences to unfiltered human connections. At the height of my drinking, I was working an unstimulating nine-to-five office job. I felt like a bull in a chute, poked and prodded and raring to go with a power plant of bound energy and yearning and no outlet for it. Alcohol would open the chute. My weekend drinking gave me permission to be wild and free in an increasingly sanitary, orderly, rule-based world.

Each night came with opportunity and unpredictable rewards. Would I end up at a party where I'd have the liquid courage to speak to a girl I thought seemed nice? Would I end up closing down a bar and belting out "Friends in Low Places" with a group of new people I'd otherwise never associate with? Would I write something compelling I otherwise wouldn't have thought of sober? Would I wake up with hilarious stories to laugh about for years?

I'm not alone. Scientists now know that the scarcity loop is a crucial reason substance use continues despite adverse consequences. Researchers at the University of Waterloo wrote that the unpredictable rewards of substance use "make the behavior even more resistant to . . . [adverse consequences] and extinction." Consider illegal drug use. Much of the rush people get is from the loop that comes from acquiring and using drugs. Will we find drugs? Will we get arrested during the deal? How much of the drug will we find, and how strong will it be?

Studies show that using drugs of unpredictable strengths at unpredictable intervals—essentially any street drug use—leads to greater abuse rates. It makes use more exciting and even leads to higher highs.

It compels us to do it again and again, thanks to the scarcity loop. Without the loop, addiction rates plummet. This is one reason medically dosed drugs have much lower addiction rates. For example, a review of the research found that just 0.27 percent of patients who were prescribed opioids by their doctor developed signs of addiction. The dose is always administered at the same strength and at predictable intervals. There's no gamble. It's why recovering heroin addicts who visit methadone clinics don't usually get high from the drug.

Hence, the National Academy of Sciences found that "there is little evidence that decriminalization of marijuana use necessarily leads to a substantial increase in marijuana use." The scientists said this is due to a diminished "forbidden fruit" effect. Eating forbidden fruit is far more exciting than unforbidden fruit.

Another old study from the U.S. Department of Justice found that marijuana use seemed to fall in some states that decriminalized the drug. The experts believe this phenomenon is in part because there's no game: rewards become predictable. Other research shows alcoholism rose 300 percent during Prohibition, when the government made alcohol illegal from 1920 to 1933. We still celebrate some of the scarcity loop excitement of bootlegging with NASCAR, a sport that evolved from bootleggers who souped up their vehicles to outrun authorities.

My drinking opened a door to new, fun experiences because it permitted me to be a new, fun me. Once I started drinking, unpredictable positive rewards led me to believe that having another drink was the most rational decision. But pushing the edge more and more over time began creating long-term problems. I got arrested while trying to set a land speed record on a collapsible scooter. I ended up in a hospital after breaking some ribs one wild night. I couldn't find my car one day. My relationships went to hell. Eventually, I wanted to stop, but nothing in my life was changing. My Monday through Friday was still an unfulfilling pressure cooker. The gray feelings of restlessness and discontent built and built until again I would drink to reclaim that open and infinite sense of possibility, persisting despite negative consequences. Because maybe, just maybe, things would be great again this time. So why not play the game and spin the reels?

Some new thinkers have argued that addiction is a social phenomenon. "The opposite of addiction is connection," they argue. But a lack of connection doesn't explain addiction for everyone. Many addicts report feeling connected to others and have a strong social support network. I felt that way. Some are missing something else.

The missing piece may relate to what Zentall described as the optimal stimulation model. "It says that animals and we humans have a level of stimulation that we prefer," he said. "And when it gets below that, we search for stimulation." As he explained, "When you think of humans today, I think a lot of us get bored with how easy it is to get resources. We spend less of our time exploring and foraging for food. We spend less time outside. Our social worlds have changed. So we search for other ways to fill this gap in stimulation, to distract or comfort ourselves. . . . We gamble, we shop online, we eat just to eat, we overuse social media, or even do drugs."

The hundreds of sober people I've spoken to all say that alcohol or drugs initially served a valuable purpose for them, too. It allowed them to feel and behave like the people they wanted to be. Or it killed boredom or helped them work harder. Or it numbed their anxiety and lowered their inhibitions. Or it allowed them to escape from some restless passenger inside them they couldn't quite understand. And then the phrase is always "It worked until it didn't."

People in the grip of addiction are like Dante's image of Satan in the *Inferno*. Dante describes hell not as a world of heat and fire but as a world of cold and ice. Satan wants to get out of hell and do the right thing. All Satan's life, flapping his wings has worked to get him somewhere, but he's now stuck in ice up to his chest. To break free, he flaps his giant wings harder and harder. But he never realizes that by flapping his wings, he's creating an icy wind that makes hell and its ice colder and firmer, securing him ever tighter in his stuck position. That's what it's like.

Addiction, in other words, is a learned behavior that once worked well but begins to backfire. Using a drug or drinking still relieves discomfort, provides stimulation, and solves problems in the short term. But it starts creating long-term problems. The more often we repeat it, the deeper we learn it, the harder it is to break. Meanwhile, the problems pile up.

"Learning influences our ability to make choices," said Szalavitz, the addiction writer and researcher. "So if you learn that this one thing really makes life worth living or solves your problems, you're going to make some rather weird choices, and it's going to look like you're hijacked. But you truly believe that this is the best choice given the context that you're in. It's a completely rational choice because, if your life sucks and this is how you can get meaning and pleasure, well, that's a rational choice."

And it can happen with anything that a person believes makes their life worth living and solves their problems in the short term. Escaping into video games, intense exercise competitions like Ironman triathlons, working to climb the corporate ladder, following or playing sports, eating, anything. And it's not a bright line so much as a spectrum. For example, the *DSM-5* lists eleven different criteria, a sort of checklist, that psychiatrists can use to figure out if and to what extent their patient has a use disorder. If a person fits two or three criteria, their case is "mild." If it's four or five, their case is "moderate." Six or more and it is "severe."

This isn't to say there isn't a brain component to addiction.

After our hour-long wait at the Baghdad police compound, Mohammed Abdullah, the head honcho, emerged from his office to call us in. Erbeed and I filed in silently and sat on an emerald-green leather couch. An overhead TV was blaring Iraqi news. A bombing, something about ISIS, a protest.

Across the room stood an oversized safe on which hung a photo of Husayn ibn Ali, the seventh-century imam and grandson of the Prophet Muhammad.

The detectives in the waiting room told questionable jokes and laughed often. Abdullah, on the other hand, was all business and brains.

He looked like the coolest character from an old episode of *Miami Vice*. He had a thick black mustache. Light gray suit. Underneath, an ultra-white button-down shirt. No tie, lots of chest.

Abdullah unbuttoned his suit jacket as he leaned back in his office chair. He first questioned my sanity for coming to this part of Baghdad. Then he made a final decision: no, there would be no ride-along because it's too dangerous and he hates paperwork. Erbeed shrugged—what's another loss when your itinerary is already zero for eight?

But Abdullah was willing to talk and show us around. He began echoing Abdul-Razaq.

Abdullah detailed why he believes Captagon is booming. "We have high unemployment. Iraqis are angry. They feel left behind and left out, but don't have an outlet," he said. "Our government doesn't allow drinking. We used to not have many drugs. But once cheap Captagon started flooding the market, use became significant."

Then he rose and led us to the holding cells. The larger cells were packed with men, all standing or sitting in what space they could find. Paramilitary guards stood with machine guns draped across their chests and sidearms attached to their legs.

Most of the prisoners were young. And that wasn't by random chance. The *DSM-5* explains that people from age eighteen to twenty-nine have the highest substance use rates. The data shows "use disorder among adults decreases in middle age, being greatest among individuals 18- to 29-years-old (16.2%) and lowest among individuals age 65 and older (1.5%)."

Research suggests that many people start "learning" addiction in their teens. From the time we hit puberty until we're about twenty-five, our brain undergoes a massive renovation. During this time, we're learning how we cope with problems and find comfort. Consider: If we wait until we're twenty-one to drink, our odds of developing an alcohol addiction are 9 percent. But if we start drinking at fourteen or younger, the odds of addiction become 50 percent. A coin flip. These same figures hold for most drugs and maladaptive behaviors.

And this brain component is not unique to drugs. "Drugs and our other addictive behaviors and substances are potent, but they're not magic," said Kent Berridge, the University of Michigan neuroscientist. PSAs tell us that using a drug just once leads to addiction. But Berridge told me to picture the following scenario: Let's say we give a group of people a powerful drug like heroin. Despite the drug spiking dopamine a thousandfold, about a third of people will hate the feeling it delivers. Hate it.

Another third will feel indifferent to it. They could take it or leave it but have other things they'd rather do. That could be another drug, or something else they find more pleasurable, like eating cake, shopping, playing video games, or exercising.

A third of those people will enjoy it. And if those people who enjoy heroin continue to use it regularly, "about 15 percent will become addicted to it," said Berridge. That figure is based on multiple well-controlled studies and surveys conducted over decades.

About 10 to 20 percent of regular users of even the most powerful drugs develop addiction. Marijuana's addiction rate is less than 10 percent. Other drugs like alcohol, cocaine, and heroin hover between 10 and 20. Tobacco has the highest rate of addiction for regular users, at 30 percent.

I noted to Abdullah that it seemed to be mostly twentysomethings in the cells. "Yes," he said. "It is. But there is no average Captagon user. We've found people using the drug as young as fifteen and as old as sixty-five."

People who become addicted later in life often discover that drugs or alcohol help them cope with a bad situation or dark horizon, Satel explained. For example, it relieves the anguish of the loss of a job, the death of a family member, or a downturn in their general living situation.

So NIDA is right. Addiction is not a choice. Instead, it's a summation of repeated choices that make a different choice harder to make for environmental, biological, and historical reasons. It's deep learning.

Much like gaining weight. Few people set out to become obese. But over time, the weight accumulates, and we find ourselves obese. It happens through small decisions we make daily. Decisions about what, why, and how much to eat multiple times a day that become habitual.

And this same pattern applies to spending too much time online, binge shopping, working too much, gambling too often, obsessing over exercise or another hobby at the expense of family time—any habit that provides short-term escape and comfort but causes long-term problems.

"The problem isn't the substance or behavior of choice," Szalavitz told me. "The problem is why you need that drug, why those drugs appeal to you, and why you are trying to get out? What are you trying to escape? What do you need in your life to feel comfortable and safe and productive?"

Just as a person can learn addiction, they can also learn better coping methods. "Dark horizons can be brightened," Satel told me.

Take me. I couldn't change my issues ranging from restlessness to social situations to general life dissatisfaction that led me to drink. I could face them head-on, no matter how uncomfortable, and find other ways to manage. I could seek different ways to find the psychological state where anything felt possible. It took me a while to get this. I did a lot of wing flapping, but eventually I realized that I had to do a lot of difficult work to melt the ice.

I asked for help and worked to understand and find other outlets for my feelings of restlessness and discontent. I accepted that I don't always have to be comfortable. And life got better. By refusing to make one choice, I had to pick another. Once we realize and accept that flapping our wings only makes the situation worse, we can find other ways out of hell.

The neuroendocrinologist Robert Sapolsky remarked to me, "The fact that most people don't get addicted to addictive substances [despite their powerful effect on brain chemicals] or, if they do, eventually stop using is fascinating." And he's right. One in ten Americans say they've gotten over a prior drug or alcohol problem. Half of those people got over it on their own. In any given year, a person with a substance abuse issue has about a 15 percent chance of quitting. The years pile up, and by age thirty half of people who report problematic substance use have quit.

"There's a zillion ways a person can recover," said Szalavitz. "But they tend to all be getting a new passion and a new sense of meaning and purpose and community and connection. One person might fall in love with exercise. Another might get really into rock climbing. Another person might hate both of those things and need to see a specialist for an issue like childhood trauma they were managing with drugs. But it's all about figuring out how to find your way in life again and deal with whatever led a person to become addicted in the first place."

Today the children of the wars in Iraq are in their teens, twenties, and thirties. Syria is flooding Iraq with Captagon. A confluence of pain, product, and few ways to find positive stimulations and manage discomfort—all at a time when many are seeking comfort—is pushing Captagon use into staggering heights. And not just in Iraq—among the bored young and wealthy people and overworked immigrants in the oil-rich Middle Eastern Gulf countries as well.

This is the story of substance abuse, the most pernicious scarcity loop. The United States' opioid epidemic hit the Midwest and Appalachia hardest. These areas were once manufacturing hubs, but outsourcing left many jobless and hopeless. Opioids were abundantly available on the street. Use, in turn, boomed to comfort people from the dark horizon.

Morphine production spiked during the Civil War to treat injured soldiers. But after the war, rates of opioid abuse were significantly higher in the South among whites. Among southern blacks, meanwhile, rates of opioid abuse plummeted. One group was defeated, another freed.

And now the United States is facing a new drug problem, similar to Captagon in Iraq. After the government relaxed regulations around telemedicine prescriptions during the pandemic, questionable online prescribers popped up, and use of the stimulant Adderall surged. Prescriptions have tripled since 2009, and today one in eight Americans is using the drug. Some of these are warranted, but many aren't. The DEA worries we're now headed into a new chapter of drug addiction. To supply the demand for pills, fake prescription pills are pouring in from Mexico. And they're dangerous. In 2022, the DEA seized fifty million fake pills tainted with fentanyl. We, too, have billions of pills circulating.

"When communal conditions are dire and drugs are easy to get, epidemics can blossom," wrote Satel. In the United States, we refer to the casualties of this type of addiction as "deaths of despair."

"You have to eradicate the want—why people want to use—or you will always have drug problems," one of Satel's patients in Ironton told her.

Dr. Abdul-Razaq seemed to agree. As we rose to leave his office and return to Baghdad's overwhelming heat and hustle, I asked him one last question.

"How do you help patients who come to you with addictions or even compulsions around other habits?"

"My main advice is to make a big change," he said. "Change your circle. Go to school. Educate yourself. Get a job or change your job. Take courses to improve your skills. Learn to read and pour yourself into books. Actively go out and make friends or change your friend group. Make big changes." Embrace short-term discomfort to find a long-term benefit.

This leverages the first and second ways to get out of a scarcity loop. First, it takes the opportunity away. When we make a big change, new and more enticing opportunities can open up, and those come with their own set of unpredictable rewards. And that alters the second phase. These new rewards outweigh the short-term comfort of substance use. Szalavitz explained that a key reason most people get clean by thirty is that "once people hit that age, suddenly their life is more constricted. They don't have endless time to sit around and get high or recover from hangovers. They get married and get a career and start raising kids, so they begin finding meaning and purpose and love and comfort elsewhere."

The pandemic accelerated want, leading people to seek comfort. Entire generations learned unnatural new ways to find stimulation and cope. "The pandemic was an enormous stressor on society," Szalavitz told me. "I think people who were at the edge often went over it. I think this stress is why we're seeing many people starting to do all kinds of dysfunctional things. Like, people are going to need to do *something.*"

But the pandemic was just a powerful push toward a strange new kind of living. A type of living that compels us to seek stimulation from, yeah, *something.*

We are increasingly living like Zentall's pigeons, cooped up in dull cages unlike the wild and highly stimulating environments we evolved in. And like the pigeons, we're more likely to fall into the scarcity loop. As technologies leveraging the loop push us into more, faster, and stronger, our problems will become more significant. The strength of new lab-based drugs like fentanyl has propelled overdose deaths to record rates. They rose 33 percent since 2020.

The lingering question is what happens next. Will we be like the soldiers who were airlifted out of Vietnam? Or will our new habits last?

Just as problematic substance use thrives off the scarcity loop, so do many other behaviors we want to break out of. But each of these behaviors comes with its own set of unique challenges—and solutions.

Problematic use of new technology, in particular, has recently boomed. Especially the technology leveraging the scarcity loop. The data shows just over 2 percent of the world's population is addicted to drugs or alcohol. A study in the *Journal of Psychiatric Research* suggests the percentage of people addicted to technology is now roughly the same.

This is why the American Psychological Association is now recognizing tech use disorders and scrambling to treat them. But the answer isn't to use our phones less. Not even close.

"Like with drugs, problematic technology use is never just about the product," Nir Eyal told me. He's a former Stanford professor and tech entrepreneur who wrote the book *Hooked: How to Build Habit-Forming Products*. He continued, "Problematic technology use is about the interaction between the product, the person, and their ability to cope with discomfort, and then a situation in their life that causes a pain that they're not equipped to deal with. When you have the confluence of these three factors at the same time, it can create an experience you can put yourself into where tech products help you immediately forget about all your problems."

When I returned from Iraq, I wanted to learn more about that. What is it about tech products and the scarcity loop that can be a comforting escape in the first place? My questions led me to a thinker who is realizing that the numbers, data, and figures embedded in many uses of the loop can be their own sort of drug. But the ways all these numbers influence our behavior—not just in tech, but also some of our most common behaviors and critical institutions—are striking.

Certainty

The fungus *Ophiocordyceps unilateralis* seems innocuous at first. When it enters a tropical carpenter ant's body, it floats around the bloodstream. But the fungus slowly and quietly begins to multiply. The ant is unaware of this. Eventually, however, the fungal cells reach a tipping point. They all interconnect with small tubes and create a vast network inside the ant's body. This is when the fungus begins to play puppet master.

The fungus cuts off communication from the ant's brain to its muscles. Then it secretes chemical compounds that invade the ant's mind, taking over its behavior. The ant effectively becomes a zombie whose body acts as a capsule, protecting the puppet master fungus while doing its bidding.

The ant is now in what scientists at Penn State call "a death grip." The fungus eventually walks the ant up a grass stalk about ten inches above the jungle floor, ten inches above the ant's uninfected ant friends. This height is the perfect temperature and humidity for the fungus to grow even more powerful.

Then it forces the ant to bite on a leaf and hang. Inside the ant's body the fungus grows spores that rain down on the ant colony below—

infecting other ants. Entering their minds and altering their behavior, creating a vast social network of zombie ants.

Tulie Bakery in Salt Lake City caters to the University of Utah hipsters willing to pay too much money for a croissant and coffee sold in what looks to be a warehouse designed by Marie Kondo. The place's exterior and interior walls and ceilings are stark white. Its floor varnished cement. Air ducts exposed. Minimalist.

I was there to meet Thi Nguyen, a professor in the philosophy department at the University of Utah. I'd read a few of his sweeping philosophical papers. His ideas about how numbers are changing us echo the scarcity loop.

Nguyen wandered up looking as if he'd just come from a rock-climbing crag or a Phish concert. He was wearing all Patagonia everything. A jacket covered a sweater anchored above loose pants. His long, messily bunned black hair spilled from underneath an oversized beanie.

As we approached the counter to order, Nguyen got very intentional. He leaned forward, hands to knees, and squinted into the glass case of baked goods. "Can I please have a chocolate croissant, an almond croissant, a morning bun, a coffee cake, and a . . . ," he said as the counter attendant removed baseball-mitt-sized glistening croissants and brick-like powdery cakes.

"And let's also do a tea cake and . . ." He paused. His eyes scanned the pastries. "You know what . . . that's enough." Nguyen was once the food critic for the *Los Angeles Times*. And so this order, like any order, was serious business.

Nguyen noticed my expression—equally impressed and concerned about the four thousand or so calories of flour, butter, and sugar he'd requested. He addressed the aggressive order. "When my family heard I was meeting you here, they gave me a list," he said.

With coffee and carbs in hand, we sat in chairs surrounding an outdoor table along Fifteenth Street. It had rained earlier in the day, and the sky was all psychedelic gray clouds. A cool mist rose from the pavement and landscaping and carried the smell of mulch.

The gig at the *L.A. Times* was happenstance. "I was always really into food, and I used to drunk post restaurant reviews online," he told me. "An *L.A. Times* editor came across my posts and was like, 'Do you want a job?'" That gig helped Nguyen pay the bills as he studied for a PhD in philosophy at UCLA.

"So," Nguyen said, "why did you want to talk to me?"

Nguyen has become a contrarian thinker on the modern trend of gamification. It's when we try to insert gamelike features into everyday human behaviors and institutions. I wanted to talk to him about that. And numbers. How the rise of numbers, data, and figures are changing our brains and behavior. How they're being used to gamify our world and influence us and what that's doing to us.

He nodded and asked me what got me thinking about this. And this, as it were, led to an entirely embarrassing admission.

It all started with Instagram, I explained. I first joined Instagram, as most people do, to share photos with friends and family. "Like, a photo of me skiing with my college friends. Or at a concert with my wife. And here are seventeen different photos of my dogs," I said. "But then that all changed."

I told him I began noticing which photos were getting the most likes and comments and recruiting new followers. "And these followers weren't just old friends," I said. "They were people I'd never met. I had no clue who they were. So I began posting more of the kinds of photos that boosted my metrics. Stuff like my runs in the desert, or work travels, or ..." I named some other examples as he munched on the almond croissant, leaning forward so the pastry's powdered sugar and buttery flakes snowed down onto the table and not his clothes.

I was telling him about my Instagram habit in a whisper. It felt shameful to talk about my desire to be "liked."

"But the bigger problem with this is that I found the Instagram app creeping into my thoughts and changing my experiences and actions," I said. "For example, I do a lot of trail running. And trail running was my zone-out time. A time to completely clear my mind. Kind of like a moving meditation. But once my run photos started racking up likes, comments, and followers, I found that my thoughts while running changed. It killed the meditative aspect of running. Instead of being present and unwinding, I'd be scanning for the right scene the whole time. I'd turn a bend in the trail and think, wow, that would be a great photo for Insta . . ." Nguyen dropped his croissant. A small storm of powdered sugar billowed out upon impact. He cut me off.

And this is when the conversation turned into a group therapy session.

"My experience is something similar on Twitter," he said. "The first time or two I had a tweet go viral on Twitter, I found that I was starting to view the world through what would make a funny tweet. I would be walking around and thinking, but instead of following a thought into something richer and deeper like I'd normally do, my mind would focus on figuring out what a tweet-sized, funny thing I could produce around that thought that might go viral," he said. "So Twitter was getting into my brain and how I think, seizing my motivation systems."

And of course, we both said, we'd then get sucked into a vortex of refreshing and scrolling the app.

Nguyen started thinking about how numbers and gamification influence us while studying for his PhD. Board games ignited his interest. We're not talking about Candy Land or Battleship. In the late 1990s and early 2000s, board games underwent an intellectual revolution.

"There was a new wave of European board games," said Nguyen. "They were so well designed and so mechanically innovative. You'd be building these complex, interesting market structures out of simple rules. You'd have to manipulate fluctuating markets, and it was just so fascinating."

Take the game Catan, which has sold more than thirty million copies worldwide. Players take on the role of settlers and the goal is to build a complex network of cities and acquire and trade resources.

Some games even stood as profound think pieces. In the game Train, for example, players build and optimize a railway network. But as you get deeper into play, the game reveals that it is set in Nazi Germany and the railway network is moving people to concentration camps. The game's designer, Brenda Romero, wanted to force players to decide if they'd stop playing or continue to win. Train is more of an art piece not intended to be played more than once, but it exemplifies the profound thinking occurring in the genre.

"So I started thinking and writing about games and reading academic stuff about them," said Nguyen. But all the academic literature he read, he discovered, was entirely backward. It viewed games as a way of telling stories. Like my experience trying to figure out slot machines, Nguyen needed to skip the academics and follow the money.

"But this didn't feel at all like why I loved games or how game designers thought about games. None of the best game developers said anything about telling stories," Nguyen told me. "The best game developers were all saying that the most important tool in their tool kit was the point system. Because the point system sets the player's motivations. It sculpts what you care about. What makes games work is point systems that are incredibly narrow, incredibly simple, and incredibly crisp."

The point of a game is to enter into a small world that is an escape from our everyday life. It's a challenging but captivating diversion. We can't predict if we'll win. We play games, as the philosopher Bernard Suits explained, to voluntarily take on unnecessary obstacles for the sake of *maybe* overcoming them. The scoring system tells us precisely what unnecessary thing we must do to win.

And not just in board games. In all games and sports. From marathons to climbing a mountain to football. The point of these games isn't to arrive at a point in space 26.2 miles from where you started. Or to stand on top of a peak or in an end zone while holding a football. The point is to go through a set of clear and artificial rules that make reaching the goal a righteous pain in the ass. So much so that the reward of a win is unpredictable.

Then we struggle to end up at the finish line, on the summit, or in the end zone. If the point wasn't to face the uncertain rewards provided by these pain-in-the-ass obstacles, we could take an Uber to the finish line, a helicopter to the peak, or have everyone grab a practice ball from the sideline and walk it over to the end zone.

Nguyen wondered why that is. Why do we put ourselves into these fabricated struggles that, in the grand scheme of things, give us problems for the sake of having and (maybe) overcoming them?

The answer is complex, but it has to do with the scarcity loop and how it can be a fun and comforting escape from the chaos of modern life.

Modern life is, indeed, chaotic. A tsunami of information to consider and choices to make hits us every day. We have a million different things we could do, reasons we could do them, and decisions we could make along the way.

This is different from the human experience for most of time. As ancient humans evolved, life was simpler. We all mostly did the same work together. We'd hunt or gather. We'd make tools and build shelters together. We made decisions based on the same knowledge passed down from generation to generation. Human wisdom was like a shallow pond that everyone drew from.

But that's changed. "One of the biggest transformations in my life as a philosopher and how I think about the value of games came from a book by the philosopher Elijah Millgram called *The Great Endarkenment*," said Nguyen. "Millgram argues that the problem of our era is information overload and having experts who are extremely hyperspecialized."

Millgram pointed to the rise of hyper-specialized experts and career fields. He wrote, "More and more, we're dependent not just on other people, but on differently specialized other people. You can literally no longer understand specialists in other fields; they work to standards you're unable to make sense of, and which in any case aren't *your* standards. . . . This is true of everyone: experts rely in just this way on experts in other fields. You can no longer decide what to think or to do by yourself; questions having to do with autonomy . . . are simply moot."

Human wisdom is no longer a shallow pond. Now it's akin to a multitude of bottomless lakes.

"So now the essential problem people face," said Nguyen, "is which experts and information do I trust and how do I figure out how and why to trust them? No person can know everything. You also can't figure out the right experts. So in life, we have to do this painful and awkward thing of trusting people and information and making decisions beyond our understanding."

In everyday life, we must make big and small decisions based on imperfect information. Like, how *do* you know you did the right thing? How do you know you married the right person? Or took the right career path? Or bought the right car? Or took the right job? Or raised your kid right? Or ate the right food to avoid disease? Or worded that email to your boss in a way that helped your career?

Take a few moments to consider all of those questions right now. Anxious? Me too!

Nguyen is right that this uncertainty is anxiety inducing. A study in *Nature Communications* discovered that humans hate uncertainty so much that we'd rather experience punishment. The scientists had participants play a computer game where they clicked rocks on the screen. Once you clicked a rock, it would overturn and reveal whether a virtual snake was hiding underneath. The participants received a painful electric shock if the rock hid a snake. So it went like this: Click. No snake, no zap. Click. Big snake, big zap.

As the participants clicked, their stress levels were monitored a few different ways. By how sweaty their skin got, by how much their pupils dilated, and by constantly asking them how stressed they felt. The scientists also, quite evilly, continuously changed the pattern and percentage of snakes along the way. This made the task an unpredictable nightmare.

The findings: The players were most stressed when they were most uncertain whether their click would lead to a shock. They became the sweatiest. Their eyes the most saucerlike.

But the wildest and most important part of this study was this: When participants were more confident they'd be painfully shocked, they were *less* stressed than when they felt as if getting shocked were a coin flip. Knowing we're going to get punished is less stressful than not knowing. Hence the phrase "just get it over with."

And this makes good evolutionary sense, according to the neuroscientist Marc Lewis. This phenomenon used to help us survive. For example, if we were hunting and had a sure kill on our hands, we wouldn't get too stressed out and, in turn, work too hard. In that scenario, the extreme actions and fight-or-flight effort the stress caused would be for nothing and waste energy. The same goes if hunting the animal seemed to be a long shot. Our stress system wouldn't react as powerfully, so we wouldn't waste the effort. But our stress system would go wild, and we would put in enormous effort if we believed we had a fifty-fifty shot at the kill.

The best scarcity loop designers know this. Slot machines, for example, have the optimal schedule of losses disguised as wins. The shift from analog to digital machines allowed designers to optimize the loop to compel effort and quick repetition.

As we sat outside the bakery, the silver clouds moved south like shiny parts down an assembly line, eventually revealing sunlight. Nguyen pulled off his beanie, and his black hair spilled past his shoulders.

Nguyen reiterated that modern life is one big circus of uncertainty. "What we should value is unclear. And our values are hard to balance," he said. This uncertainty is stressful and uncomfortable. And it's everywhere. Always. Everyday life is often the opposite of a fun escape. Rather, it's something we frequently want to escape.

Enter games. This is why games are so powerful and why we play them. "Games are a balm for the confusion and anxiety of real life," said Nguyen. "They give us a little world where we can manage and understand everything. We know exactly what we are doing and why we are doing it. And when we are done, we know exactly how well we have done. Games offer us a momentary escape from the confusion of the world." They do this with those, as he put it, "point systems that are incredibly narrow, incredibly simple, and incredibly crisp."

Consider the slot machine. Each game has an unpredictable outcome, but there is zero uncertainty over the outcome once the reels fall. We know *exactly* how we did. We either lost our money, won less than our bet, or won more. That's it. That's the entire game.

Which brings us to gamification.

In 2010, the researcher Jane McGonigal gave a viral TED talk on games. She was, at the time, the director of game research and development at the Institute for the Future, a Silicon Valley think tank. In her talk, titled "Gaming Can Make a Better World," she argued that we should be using the mechanics of games to alter human behavior. She said we should "gamify" the world.

"When we're in game worlds," she said, "I believe we're the best version of ourselves. . . . We're most likely to stick with a problem as long as it takes, to get up after failure and try again."

It's debatable whether we're the best version of ourselves in game worlds. McGonigal seems to have never encountered the term "sore loser." Or stood at a craps table with a gambler chasing losses. Or watched two NHL enforcers punch each other's teeth in. But she's right that games compel us to stick with it, get up after failure, and try again—often for too long. That's the scarcity loop.

McGonigal said this was different from how we face most problems in life. "In real life, when we face failure, when we confront obstacles, we often don't [get up and try again]. We feel overcome, we feel overwhelmed, we feel anxious, maybe depressed, frustrated, or cynical. We never have those feelings when we're playing games; they just don't exist in games," she said.

The reason: Games thrust us into a scarcity loop and give us a score. The score gives us anxiety-relieving certainty about whether we won or lost, or did the right or wrong thing. Games incentivize quick repetition, or getting up and trying again.

McGonigal argued that by gamifying the world, we could "build up bonds and trust and cooperation" and find "blissful productivity" and "epic meaning." She thought this could solve our big personal and societal problems.

McGonigal had even conducted a series of studies on this idea. One, called World Without Oil, put seventeen hundred players in a virtual world of oil shortages. It gave the players real-time information about how those shortages affected food supply, transportation, school closures, rioting, and more. Players had to figure out how they'd use less oil. But the critical takeaway, McGonigal said, is that many players also altered their oil use in real life.

It's a fascinating finding with potential for good. But it wasn't just save-the-world types who watched the viral TED video. It was also companies across industries.

Within a few years after the TED talk, more than 350 major companies launched gamification projects. MLB, Adobe, NBC, Walgreens, Ford, Southwest, eBay, Nike, Panera, Starbucks, Oracle, and more.

They all pulled from elements of the scarcity loop. Points, leaderboards, badges, performance graphs, constant notifications, and more started to appear in all kinds of digital spaces to bend our behaviors. They offered opportunities, unpredictable rewards, and quick repeatability. We began gamifying anything we could: exercise, learning, weight loss, shopping, advertising, and even health care and warfare.

The U.S. Army created a shooter game called *America's Army* to recruit soldiers. Mars candy made an I spy game to sell pretzel M&M's. NikeFuel pitted users against each other to see who could rack up more exercise for virtual trophies and discounts to share on social media. The tech company Bluewolf launched a game for its employees that awarded points and rewards for sharing content about the company on their personal LinkedIn and Twitter accounts.

The gamification industry went from nothing to $2 billion in a few years. Companies that had long struggled to shape our behavior discovered something intoxicating. If they just gave customers or employees a clear goal—"a little world where we can manage and understand everything," as Nguyen put it—we will be more likely to do what they want us to. Companies saw an average 30 percent increase in how much time people spent on their sites, revisited them, and shared content about their brands.

"So the idea with the gamification movement was that the more you make life a game, the more awesome life will be. But," Nguyen said, "that is incredibly naive."

Nguyen took another bite of the croissant and began explaining how he arrived at this conclusion. "If you look at regular games like a board game or sports game, the actions are screened off from ordinary life," he said.

In regular games, it doesn't matter that the goal is to abide by silly rules in the name of arbitrary points because regular games happen *outside* everyday life. The *entire point* is to abide by silly rules for stupid points for a screened-off moment. Remember we play games for a fun escape from everyday life. The slot machine stays in the casino (or gas station or grocery store). The marathon ends at the finish line. Clue ends when we discover it was Colonel Mustard in the study with the candlestick. So do the repercussions: Our significant other doesn't *actually* care that we financially devastated them in Monopoly. Football players don't tackle each other off the field. If they do, they wind up in jail.

"But when we try to gamify ordinary life," said Nguyen, "we're trying to impose clear values on a preexisting thicket of values, on a system that is very uncertain and complex." We abide by silly rules in the name of arbitrary points for parts of life that *aren't* silly and arbitrary.

Nguyen continued: "So instead of figuring out why you care about something, how you care about it, and articulating for yourself what it is about that system that's important to you, gamification just tells you, 'Here's how and why you care about it.'" And we fall in line. "So gamification can increase motivation, yes, but at the cost of changing our goals in problematic ways." And the repercussions ripple into actual life.

"The first place I started thinking about how numbers shape our behavior is with wine scoring," Nguyen told me.

Until the 1970s, wine critics were all a bunch of cork dorks who described wine in a way that was an indecipherable adjective-and-noun salad. They'd write about a wine being, as one reviewer did, "tightly focused, with a beam of linzer torte, bitter cherry, plum sauce and fig fruit laced with licorice snap, singed iron and roasted bay leaf. The long finish has lots of roasted fig, tar and spice notes for extra bass, but the acidity is there as well, deeply embedded."

This pomp made the average person wonder how deeply most wine critics' heads were embedded up their behinds. But in 1979, the wine critic Robert Parker changed everything. Parker was a blue-collar guy from the woods of northern Maryland. He loved wine and wanted to sell it but had no formal training. The current state of wine criticism, he thought, was keeping wine from getting into the hands of people who would enjoy it but were turned off and confused by the snobby language.

So, long before McGonigal gave her talk, he gamified the system. He began ranking wines on a scale from 50 to 100 points. He published those reviews in his magazine, *The Wine Advocate.*

Parker's system quickly and single-handedly gamified the wine-making industry. Thanks to the clarity of scoring, wines with a score of 90 or above flew off the shelves while bottles scoring below 80 collected dust. One wine shipper in Bordeaux told the journalist Elin McCoy that winemakers would make $7 million more off a single wine if it scored a 95 instead of an 85. If Parker gave a wine a 100, a vintner could increase prices by 400 percent.

Wine geeks called what happened next the "Parkerization" of wine. Parker's opinion and palate, explained one wine magazine, "became the only one that mattered to collectors as well as winemakers." Winemakers began changing a wine-making tradition that had existed for thousands of years. They altered how they made their wine to suit Parker's tastes to bump up their scores and sales. Parker's system was so popular that other critics also began adopting scoring systems.

But what if you don't have Parker's tastes? Then the scores are meaningless.

"And part of what makes a wine good is that it pairs with food and reacts differently with different food," said Nguyen. A wine might be too powerful or subtle on its own, but when you pair it with food, it expands or explodes, and it even changes over the course of a meal,"

He continued: "In order to render a point score not just with wine but with anything, you need to be objective. You can't have the wine with food, because that's too much of a variable. So the wine-scoring world has small glasses all next to each other. And that eliminates the relationship with food and dynamism and the *most important* thing about wine in the first place." And once you spend four times more money on that 100-point bottle and pair it with your dinner, its score becomes irrelevant.

This scoring can even alter our expectations and experience. Researchers at the University of Bordeaux gathered fifty-four students in their wine studies department. They got two bottles of wine: one the students knew was cheap and another they knew was expensive. Without the students' knowledge, the researchers switched the bottles' labels.

The students, in turn, waxed poetic about the cheap wine they believed was expensive with all kinds of noun-and-adjective word salads. Meanwhile, they described the costly wine they believed was cheap as something akin to cat piss.

Nguyen's experience on Twitter—which is the same I had on Instagram, which is the same you've surely had *somewhere*—can help us understand the downsides of gamification and quantification.

Nguyen joined Twitter during a period of peak gamification. The world was moving to a gamified future. "And the first time I had a tweet that got retweeted more than a thousand times, my brain just went, 'YES. That feels good. Give me MORE of that,'" he said. "And I could immediately see, wow, this is really powerful. This is dangerous shit."

Dangerous because there was a big disconnect. Twitter billed itself as a forum for online discussion. But the points Nguyen was racking up in the form of likes and retweets didn't feel like discussion.

"The ordinary goals of discussion are complicated and complex," Nguyen told me as a slight breeze lifted the brown paper bags our pastries came in, which were now acting as crumb-covered place mats. Nguyen's crumbs slid onto the table; mine slid all down my front. As I awkwardly brushed myself off, he threw out a few examples of a discussion's possible goals. Its goal might be to understand each other and be understood. Or to share empathy. Or to commiserate. Or any number of other things. And its goals can be many of those things at once.

But Twitter's scoring mechanism replaces all those values with a much simpler goal: to rack up likes, retweets, and followers. The goal becomes for us, as it did for the winemakers, to score more points. And we, like the wine drinkers and producers, are deeply influenced by those points. Points and gamification begin to remodel our experience, our behavior, and how we define success.

"When you substantially change the goals of the activity, that changes the activity itself," said Nguyen. His tweets that racked up the likes and retweets were at odds with discussion. They had to be short and sharp and context free. And maybe even a bit dickish.

Research shows that tweets with strong moral emotions, like outrage, score the most likes, retweets, and followers. The scientists call this "moral contagion." And, just like a virus, it is indeed contagious.

This is important even for those who don't use social media. Take how it's shifted politics. In 2022, a team of researchers from Canada, Stanford, and the University of Illinois noticed that America's elected civil officials seemed to be using Twitter in an increasingly uncivil way. So they pulled a decade's worth of tweets from members of Congress. There were 1.3 million tweets total. The scientists then used an AI algorithm to analyze each of the 1.3 million tweets and assign it a "toxicity score" from 0 to 100.

The finding: Political Twitter became 23 percent more toxic across the decade. And this is because, the researchers wrote, "uncivil tweets tended to receive more approval and attention [measured by] large quantities of 'likes' and 'retweets.'" The scientists also noted that once the politicians got a rush of likes and retweets from a mean tweet, they became more likely to boost their future tweets' meanness. This scarcity loop of constant unpredictable rewards and quick feedback ramped up their Twitter use and turned them into monsters.

Politicians from both sides of the aisle fell into this nasty loop. And it might be one thing if this crappy behavior, like in regular games, were screened off from every other part of their life. But, critically, the repercussions rippled. The politicians used the same tactics to guide their policy thinking and decision making—that is, the legislation that affects our lives. Right now.

Nguyen calls this phenomenon "value capture." When we stamp a simplified scoring system on an activity, we begin to fixate on the scoring system and chase points rather than experience the activity's original goals. "Those metrics take over our motivations," said Nguyen. "We get satisfaction in exchange for shifting our goals along engineered lines but risk losing sight of the real importance of the activity. It bends toward something much more impoverished."

This impoverished arc is a fun escape that brings us the comfort of certainty through points. But it bends us away from truer, richer experiences and the real goals of the activity.

And it's not just our politicians who are chasing the Twitter high. It's shaped what we read and see and how we understand the world.

In 2022, the executive editor of the *New York Times* told his journalists to spend less time on Twitter. The reason: Since the rise of Twitter, journalists were increasingly framing their stories not in a way that found balance, truth, and objectivity. They were framing their stories in a way that scored points with their Twitter followers, many of whom were fellow journalists rather than the public. This in-group point scoring was diluting the *Times*'s worldview. The public intellectual Bari Weiss, who worked at the *Times,* called Twitter the *Times*'s "ultimate editor."

I asked Nguyen if he thinks we can use social media tools in a way that doesn't, as he put it, impoverish us. Are they just a tool in the sense that a hammer or guillotine is a tool?

He told me it depends on why and how we use it or any other gamified scarcity loop. To use it well, we would need to resist the pull of points and establish different goals.

This leverages the second way to get out of a scarcity loop. The rewards stop trickling in because we no longer find the scoring metrics of social media rewarding. We establish unpredictable rewards on our terms.

Nguyen compared this to how we can have different relationships with money. Some people see money as an end, while others see it as a means to an end. The first person works for money for the sake of it and makes decisions that help it pile up. For example, they'd always take the job with the highest salary.

The second person sees money as a tool for something else. For example, happiness. They only take an opportunity for more money so long as it makes them happier. If faced with two job offers, this person would accept the one they believed would make them happier even if it had a lower salary.

Applying this logic to social media, we'd establish a goal or two and use it only to advance that goal. Numbers and scoring would be useful only insofar as they advanced our goal. So, for example, we might use it to spread useful information about a topic we're an expert in, make new friends, or learn about different cultures. If we're using it for reasons other than that, our head might be underneath the guillotine blade.

I'd finished my raisin croissant. It was exquisite. And expansive. I felt as if I needed a nap. And then a long run to burn off the multiverse of calories it contained.

While in that state of sugar-fat-flour-induced euphoria, I was realizing that this scene—two comfortable men eating $6 pastries and drinking $5 cups of coffee and complaining about social media—is sort of also impoverished?

As Nguyen talked, I was inside my head, weighing various viewpoints. Perhaps we are entitled men-children who should grow up and worry about something else. And maybe there's something to be said for that.

But perhaps we're not alone in our experience. And maybe that experience stands for something much larger. I don't think it lives only in the world of wine and Twitter and Instagram.

And so I asked Nguyen as much. "Where else do you see scoring systems changing our behaviors and goals and experiences?" He looked at me as if I'd just risen from a thirty-year coma.

"Ummm," he said, chuckling. "Well, where do I start? We prefer metrics. So our attention and values naturally shift toward what's easily measurable at scale and away from all these other, more complicated things that matter more." It turns out he teaches an entire class on one way this happens.

"My students' favorite day in my intro to philosophy course is called 'Are Grades Bullshit?'" he said. "I never see intro students more engaged and mind blown than on this day in class."

In that class, he explains how grades arose as a way not to help students improve and think better but to make administrators' lives easier. For example, grades and grade point averages allow school admissions officers or potential employers to make quicker judgments about a person's aptitude.

But we trade speed for accuracy. Over time, grades became the be-all and end-all of the education system. "So now students *obsess* over their GPA rather than focusing on whether they understand the skills and ideas they need to thrive in the job market," Nguyen said. He's tried to ungrade—to use written evaluations. But administrators won't have it.

As a professor in a university myself, I began to commiserate with Nguyen. "I've also noticed that grades often misrepresent students," I told him. "My most promising students usually don't receive an A. They're too freethinking. Or they're gritty hustlers who work forty hours a week while taking on a full load of courses. My most promising students are usually in the B to A- range.

"My students who get As, on the other hand, are more likely to be the more robotic and less creative ones. Or they have wealthy parents who pay for their schooling and have more time to study," I said.

Yet the A students are most aggressively recruited by businesses and into the top leadership positions. This reinforces a business world that increasingly favors measurement over big, creative ideas that can move the ball farther downfield. No wonder Steve Jobs, Bill Gates, and Oprah Winfrey dropped out of college.

Numbers have also captured how we view health. Our doctors ask us to be within a certain BMI, or body mass index. Using our height and weight, BMI scores us as being "normal weight," "overweight," or "obese." Like grades, it's helpful for quick and rough analysis. But the metric doesn't capture our true physical or mental health status or all the intricacies of weight that can affect our health. For example, how much of that weight is muscle, where the weight is stored, or why we might want to be at a certain body weight in the first place. This often applies to other broad health metrics, like blood pressure and blood sugar levels. And we often receive treatments or insurance premiums we may not need based on these numbers.

Or we begin exercising for health and empowerment, then get captured by activity tracking devices and daily step counts or "strain counts." Yet we never define what health means to us in the first place. And if we did, it surely wouldn't be a number on an app.

The list goes on. Researchers have found that quantification is changing everything from how we experience movies and music, thanks to the popularity of Rotten Tomatoes and Pitchfork; to sports, with the rise of moneyball analytics mostly ruining the soul and excitement of baseball; to art and the outdoors, with people flocking to museums and outdoor sites to take selfies for social media likes; to how chefs cook meals or how we design spaces to look good on Instagram or TikTok.

The technology writer Kyle Chayka calls the latter the "AirSpace." Businesses are designing themselves to appeal to social media users. The same aesthetics are now proliferating globally, largely thanks to gamified technology. Customers want to visit a place for its typical goods—and for social media photos. Chayka points out that today coffee shops in the hip part of Salt Lake City, Austin, Tokyo, or Reykjavík all have the same Instagram-friendly look. Minimalist everything with an industrial feel. Like, ironically, this bakery Nguyen and I were at.

One reason we like escaping into the scarcity loop is that there is no uncomfortable, anxiety-inducing uncertainty in it. It's safer and easier. We can know whether we did good or bad.

It's much harder to consider the effect our time in nature has on our psyche. Or how a painting makes us feel. Or why we might go to a particular restaurant, listen to an album, watch a movie, or design or experience the interior of a coffee shop and how all those things do or don't appeal to us and why.

As I drove home to Las Vegas, I thought back to a conversation with Caleb Everett, a numbers researcher and senior associate dean at the University of Miami. He discovered that written numbers are a new phenomenon. They were developed only in the last few thousand years. "We tend to think numbers are a default modus operandi for humans, but that's not really the case," Everett told me.

In our ancient past, we didn't put a number on things. We worked with rough estimations. There are still humans whom numbers haven't reached, like the Pirahã tribe in the Amazon. "The Pirahã can't count above three, and they have no words for numbers," Everett told me. Instead, they use rough approximations, like small, medium, or large.

"It's fascinating that numbers have become so powerful to us at an emotional and behavioral level," Everett said. "They seem natural to us, but they are in many ways a very unnatural thing. Because you can make the case that we're not really hardwired for them. Elaborate numbers are a historical aberration in terms of our species. But the cultural trajectory over the last thousand, and especially the last few hundred, years has led us to quantify everything and believe that quantification means absolute truth. We think that if you can quantify something, boy, you've said or accomplished something true and profound, because you have evidence. And you can rest assured and not have to think about it anymore. But it just doesn't work like that."

We shaped numbers, and numbers, in turn, shaped us. By narrowing in on one aspect of an experience, we can miss all the other more important aspects of it. The more thought-provoking and meaningful aspects. The aspects that make life worth living.

"Because if you actually spend time with data and study data and science, you can see how complicated it is and how hard it is to make sense of," Everett told me. "There are a million ways to quantify something and arguments about how best to do that quantification. For example, what variable you'll control for or not control for and what the data actually means. These are very debatable and often flawed human decisions. It's just so, so complex. There are just so many layers of gray. But when that gray data moves down to the general public, the public often view it as black and white. As fundamentally true."

Metrics, of course, aren't useless. But we need to realize that they don't account for the human experience. All the experiences we can have throughout everything we do, and how those experiences can affect us and others.

We must see them for what they are: gray oversimplified scores that can tell us a little bit, but far from everything. Outside forces can use them to pull us into a scarcity loop that is divorced from reality and our best interests, walking us into places we may not want to be. Then it's worth occasionally taking a different path. Embracing the gray and wading into uncomfortable water to figure out why we're doing something in the first place, and what all those somethings really mean to us.

Influence

When we think of influence today, we probably don't think of far-out ideas like how tools such as numbers influence us. We likely think about social media "influencers" and our influence on others. And this makes sense.

We now know that scarcity brain evolved to crave influence. Humans are social animals who evolved to vigilantly jockey for status, fret about what others think of us, and often do what we do and think as a reaction to others.

But until recently, influence was a soft science. It was hard to measure our social rank, or how we were doing compared with others, or whether our actions were good or bad. Then everyone became connected by the internet and we could quantify influence.

Like traditional drugs, our drive to influence others falls prey to more and faster and stronger. The research shows that when we start to feel as if we have an opportunity to gain status and influence, we start valuing it even more and doing more things to get it. Whether posting on social media or behaving a certain way around others, we'll see the opportunity, act, wait for unpredictable rewards, then repeat. The scarcity loop.

In 1943, the pioneering psychologist Abraham Maslow released a paper called "A Theory of Human Motivation." It laid out his famous hierarchy of needs. It states we must fulfill our basic needs before moving on to more abstract needs. So critical needs like nutrition, sleep, water, and shelter come first. Next comes safety. This includes being free from disease and having a job and some resources. Then come psychological needs like feeling as if we belong. We go up the hierarchy until we eventually reach what he called "self-actualization" needs. He described self-actualization needs as "becoming everything [we're] capable of."

It seems reasonable to suggest that humans prioritize food or safety over friendship or deep discussions about the existence of God if they're about to starve to death or be mauled by a grizzly bear. Or that we'd care less about "finding ourselves" if we're living on the street or laid up in the hospital. And most psychologists at the time agreed.

But Maslow's fourth rung of the hierarchy was a bit more controversial. It centered on what he called "esteem" needs. Maslow argued that we not only need to feel good about ourselves. We also need to feel as if *others* feel good about us. Maslow put it like this: "We have what we may call the desire for reputation or prestige, respect or esteem from other people, recognition, attention, importance or appreciation."

This at the time was a hot take. Thinkers like Sigmund Freud had dominated academic psychology until then. Freud believed we all have animalistic drives hidden deep in our unconscious. Our problems, Freud thought, come from repressing those drives or projecting them onto others.

But Maslow believed psychoanalysts like Freud had ignored that we are social creatures. And psychiatrists on the ground quietly agreed with Maslow. Their patients, they'd tell Maslow off the record, weren't coming to the therapy chair to talk about Freudian stuff, like how they had bizarre repressed sexual frustrations. Many patients were talking about their position in society. Patients had angst over social issues like not receiving some new title at work or not feeling as rich as their neighbors.

So when Maslow released his paper, the idea that we all crave influence and social status sort of flopped. In academic circles anyway.

"Psychologists didn't want to admit that status is important, because it is somewhat unsavory," Cameron Anderson, who studies power, status, and influence dynamics at UC Berkeley, told me. "People are not supposed to care about status, and a high desire for status is seen as selfish and superficial."

Yet let's all be honest here: we do care. I've considered my rank in all sorts of hierarchies more than I'd like to admit. There's my social media habit that Nguyen helped me understand. But beyond the screen, every one of us is reminded of our rank every day.

When we go to a sporting event or concert and see the "Very Important Person" section. When we drive past our neighbor's bigger house. When fake laugh at our boss' dumb joke. When we board an airline and walk past the people sitting in roomy first-class seats sipping preflight champagne.

We like to think we're immune to these status reminders. Nope. We're not.

For example, researchers at the University of Toronto and Harvard recently discovered that incidents of "air rage" were on the rise. That is, people going batshit crazy at thirty thousand feet. The scientists wanted to know why. So they studied more than a million flights and roughly four thousand cases of air rage. They assumed the uptick in incidents was entirely due to flying becoming a more hell-like experience every year, with more fees and delays and less leg room. They were wrong.

They discovered that the chances of economy passengers spinning into a fit of anger were four times higher when the plane had a first-class cabin. This status and influence cue increased the odds of an incident the same as having a nine-and-a-half-hour delay. And if the economy passengers had to board through the first-class cabin, the figure spiked to an eightfold increase.

That awkward walk past "first class" citizens subtly makes us more likely to want to destroy the class system all together. To have a mid-flight French Revolution and line up the elites at the guillotine.

So, yeah, we all care and are all affected by cues around our social rank and influence. But most of us suppress the revolutionary, Che Guevarra-like instinct this can incite. Instead, we feign that we're above the subtle status reminders we face every day. We soothe ourselves and undercut the high-status people by telling ourselves things like, "that house is a monstrosity . . . money can't buy taste." Or "the plane lands at the same time for everyone."

Because here's the more important thing: expressing that we care, ironically, is the worst thing we can do for our status. Nothing hurts our status more than letting others know we care.

So for psychological scientists, researching social status was akin to self-admission of this. That we really, really care about what others think of us. And that we are, as Aristotle put it, "political animals" whose actions are shaped by others. And this was despite other academic disciplines realizing that most of the decisions we make are, in fact, a reaction to society and fitting into it.

For example, the anthropological theory of "schismogenesis" argues that our entire sense of self and cultural identity is effectively a reaction to others. As the scientists David Graeber and David Wengrow explain the idea in their book *The Dawn of Everything*, "By any biologically meaningful standard, living humans are barely distinguishable." We all mostly look the same. We act much the same (for example, puffing our chests and raising our hands is how people everywhere show pride). We even think the same; all human language has nouns, verbs, and adjectives. And even though our music sounds different, we all like music and dance. Meals worldwide are a carbohydrate, some protein like meat or fish or beans, and some vegetables.

But our minds always look for the relatively minor differences we have with others and form our sense of self and societies around those differences. "People come to define themselves against their neighbors," Graeber and Wengrow wrote.

Scientists ignored these topics even though social rank has been embedded into humanity for a long, long time. Today, we often view ancient societies as being free from rank.

But consider the "potlatch," a practice by American Indian tribes in the coastal Northwest. To understand it, imagine you get invited to a big dinner party. But beforehand, the host asks you for your income and title at work.

Then, when you arrive at this party, your seat at the table, the food you're served, and the silverware you use to eat it with are all based on your income and title. So you, an accountant, are sitting somewhere in the middle of the table, eating New York strip steak and a baked potato off Costco china with a stainless steel knife and fork. Meanwhile, your neighbor the tech investor is eating Kobe beef with lobster risotto and using gold-plated silverware. This is upsetting. But, hey, look on the bright side. At least you aren't the public school teacher at the end of the table eating a hot dog and store-brand potato chips with your bare hands off a paper plate.

This sounds awkward. Cruel, even. But this is exactly what the potlatch did. It was a spectacular display of social ranking celebrated with a massive feast where people were seated based on rank. Their social rank determined everything from the food a person ate to the dishes and silverware they used to eat it. The leader who threw the party would even make a big speech about his influence and then overwhelm his guests with gifts as a big flex of status. Sort of like how the first-class passengers get champagne, a hot meal, and a special welcome across the in-board PA system while the rest of us get stale pretzels, a warm Diet Coke, and yelled at to keep our seat belts fastened. And the potlatch is just one strange example of the status games cultures worldwide have played throughout time.

So for five decades after Maslow's insight, scientists avoided studying status. Lest their admission that status is important affect their status. But in the early 1990s, as the United States was grappling with issues around race and gender, a psychologist at Princeton named Susan Fiske started studying stereotypes. She found that stereotypes are one way the powerful exert control over the powerless.

Her work "brought back the topic of social hierarchy more generally to social psychology," said Anderson. The focus on the underprivileged "seemed to make it acceptable to study other topics related to hierarchy, including who attains high status and why."

The research has been piling up since.

"The reason we have emotions is because they serve critical functions that throughout history have assisted with our survival and reproduction." That's according to Jessica Tracy, a professor of psychology and the director of the University of British Columbia's Emotion and Self Lab.

Tracy told me that we have a whole toolbox of emotions that depend *entirely* on other people's thoughts, feelings, or actions. Emotions like embarrassment, guilt, shame, jealousy, envy, empathy, and pride don't work in a vacuum. These emotions are tools that compel us to work with others (or not) so we can thrive in our environment. So we can retain and advance our place in society. Let's call these our "influence emotions."

Scarcity brain craves influence because the more influence we have over others, the more likely we are to survive and spread our DNA. Influence got us better mates. It increased our odds of survival in a conflict. It got us a bigger share of scarce resources. It even helped us get out of crappy menial work that burned energy. As long as humans have been around, Anderson said, influence has been like a vitamin or toxin. "People's subjective well-being, self-esteem, and mental and physical health appear to depend on the level of status they are accorded by others."

Anderson knows that, because he recently analyzed all the data on social status. We're talking hundreds and hundreds of studies going back to Maslow. He found that people with higher status are generally happier and experience fewer mental and physical health problems like anxiety, depression, and heart disease. And those health differences aren't due to the lower-status people having poorer access to health care. The same findings have been shown in countries with universal health care, where health-care access and quality are the same for everyone.

Just as scarcity brain is always unconsciously scanning for cues that resources are scarce, it also scans for cues around influence. Even seemingly insignificant differences among people—like the size of a serving of food or drink we receive relative to someone else, a person's office decor, or differences in dress—can trigger our influence emotions and stress us out.

Most of us will even take status and influence over money. One poll found that 70 percent of workers said they'd prefer a better title over a raise. They believed the better title would lead others to view them in a more respectful light and give them more influence. Not just more influence over their co-workers. More influence over everyone else in their orbit—their neighbors, family, and friends.

This led to another fascinating finding, this time from scientists at Cornell University. It's better for our physical and mental health to be a bigger fish in a smaller pond—that is, having a higher rank at a smaller company—than to be somewhere in the middle of a big, well-known company. Even if we make more money at the bigger company.

Status ponds are more important than we realize. How we feel at any given moment is surprisingly linked to the pond we're in. For example, research shows that people in the top one percentile of wealth—one percenters who make at least $600,000 a year—frequently complain of feeling poor and stretched. This is because they usually live around other one percenters. So they focus on what they don't have compared to their peers. It leads these objectively rich people to believe that they are subjectively poor.

Other research shows that our desire for influence is also at the root of violence. A large study of inner-city crime in Detroit found that the most common reason one man killed another wasn't over money, drugs, or girlfriends per se. It was over status threats. The scientists gave the example of two men fighting over access to a pool table. The fight wasn't *really* about the pool table; it was over what ceding the pool table (or girlfriend or drug corner) would do for their status.

Our drive for influence also creates anxiety. Anderson's research shows the most stressful thing that can happen to us is taking on a big public task we feel ill-prepared for. Like being asked to give a talk on a topic you feel shaky on at the last minute. It leads our bodies to open a floodgate of stress hormones.

Or, say, we see two people whispering. Our brain often assumes the two people are whispering about us, ostracizing us, and plotting our demise. Or say someone makes an ambiguous comment. We default to thinking it was an undermining dig. On the flip side, if *we* make an ambiguous comment to someone and later realize they could take it the wrong way, we believe that the person did, in fact, take it the wrong way. Then we begin mentally planning and replanning how we'll smooth the situation over.

We're usually wrong on all counts. The comment wasn't a shot at us. The person didn't take our comment the wrong way. But our brain defaults to assuming the worst. In the off chance we're right, we save our social status. And in the past, that could mean the difference between life and death.

Of course, seeking influence isn't bad. The reason we did it and still do it today is that it helped us survive and thrive. It can lead us to work hard, to be more generous, to do good and help others. But our drive for influence can also lead to vain, selfish, and destructive tendencies like overcompetitiveness, overconfidence, materialism, aggression, and general misery. This is likely why social commitment was one of three compulsions the Buddha had to overcome to reach Nirvana.

Humans lived in small bands together for nearly all of the 2.5 million years we've been around. Our tribes never grew above 150 people. Most of them were much smaller than that. We rarely met new people. And we all had the same jobs. We had to hunt, gather, make some kids, and not die. The social dynamics weren't all that dynamic, and we could do only so many things and influence only so many other people. The number of people we could influence and the kind of influence we could have over them were scarce. The social ladder was more like a step stool.

But today there are eight billion of us all interconnected. Thousands of old social step stools have united, and the social hierarchy is more like the Tower of Babel. We can all jockey for influence at any moment, editing an image of our lives and thoughts on social media. For a group of billions. Every day all day. And it's all quantified, gamified, and measured in likes, followers, DMs, retweets, impressions, views, income, and so much more.

And this has fundamentally changed us, said Tracy. Take what's happened to how we feel pride or shame. "These two emotions evolved specifically for social rank and hierarchy," she told me. "They motivate us to do or not do various behaviors that help us attain social rank and convey to others around us that we're deserving of social rank."

Pride is the emotion we feel when we evaluate ourselves positively.

"Pride feels great," said Tracy, "and as a result of it feeling great, we want more of it. So we're motivated to go out and seek more achievements or behave in more ways that help us attain pride, which is really just a way of saying we do more of the things that make us feel like we're getting social status." But Tracy's research shows that there are two types of pride. One has a bright side; the other has a dark side.

The first type is called "authentic pride." Authentic pride is good. It's the type of pride we feel when we achieve something great and *deserve* that pride. For example, closing a big deal we've worked hard on, finishing faster than we expected in a 10K race, or creating something of value. Tracy said, "We feel authentic pride even if no one sees what we've done. But if someone happens to see the great thing we did, then it feels extra good." Authentic pride boosts creativity and long-term mental and physical health and gives us lasting influence.

Then, in the shadows, is the second type of pride: "hubristic pride." Hubristic pride is undeserved pride. It's a type we project without accomplishment. It's the type of braggarts, self-aggrandizers, narcissists, and egomaniacs. It still gives us good feelings, "but the good feelings from hubristic pride come with problems," said Tracy. "Hubristic pride leads us to behave in ways where the goal is all about getting other people to appreciate us." It's driven people mad and into self-destruction.

In short, authentic pride comes from doing awesome things. Hubristic pride comes from falsely advertising ourselves.

It used to be that we could advertise our pride to only so many people. But mass media changed the game of influence. It started with print writing, then grew with radio and TV. The influence this gave us, however, was often justified. We had to have big ideas and do interesting things to gain mass influence and public media attention.

Then we got the internet and social media. Tracy explained that if we accomplish something today, "it's very easy to cast a really wide net. You can brag on social media now, and now all of your friends know about your accomplishment. And it's super tempting to do this because the greater your audience, the more accolades you get and the better you feel." There are now roughly five billion of us on social media.

"But it's a catch-22," Tracy told me. "As you increase your status by letting people know you're great, you simultaneously decrease it by letting people know you're great. And that makes you look like a jerk. The much harder but more effective way is to actually go out into the world and do great things. And then status arises naturally."

Shame has also shifted, Tracy said. Research shows that new and controversial opinions get the most traction on social media, as we learned with those tweeting politicians. But it's a delicate tightrope. If we go too far and offend too many people, the public pile on is no longer a small group of people in our tribe shunning us. It could be a group of thousands or even millions. Cancel culture is society launching a full-on blitzkrieg of shame. The rise of smartphones with cameras and an internet connection has allowed people to put shame at scale.

This exerts a particular hell on teens, whose brains are changing in such a way that social status and acceptance become more important than at any other time in life. Research from the NYU social psychologist Jonathan Haidt has shown that striving for attention on social media has led to a surge of depression, anxiety, and self-injury, especially among teen girls. "Particularly Instagram," wrote Haidt. Instagram "displaces other forms of interaction among teens, puts the size of their friend group on public display, and subjects their physical appearance to the hard metrics of likes and comment counts."

Teens understand the pernicious effects of social media. But because their social lives exist online and sociality is uniquely important to teens, they're like moths drawn to a flame.

The answer, however, isn't just to get off our phones and use the internet less. The internet is endemic. And just when we master the art of using one platform well, another will pop up leveraging our need to be liked to pull us back into a new and stronger scarcity loop. With the rise of artificial intelligence, our games often know us better than we do. They dribble out rewards on the exact schedule of unpredictability that sucks us in.

But as Nguyen said, we don't have to play the simplified games technology asks us to play. We can invent our own game with its own goals and outcomes.

More important, we must realize that our drive to be liked and influence others influences us far outside the internet. It is perhaps the biggest determinant of the course of our life and who we become. It's steered our thoughts, actions, and well-being since birth, in the real world.

Our mind is ingrained with social conceptions, compulsions, and assumptions that alter how we see the world, make judgments, and relate to others. In the past, these helped us gain or maintain influence and survive. But today they often constrain us. Scientists call these cognitive biases.

For example, a team of scientists at the University of Pennsylvania and UC Berkeley discovered that when we perform a task in front of others, we believe we're being judged far more harshly than we are. We also think people take single moments of our lives and make sweeping generalizations.

So let's say we misspeak one line of an hour-long presentation. We afterward tend to believe that everyone in the crowd is not only mentally rerunning our mess-up over and over (just as we are!) but also thinking that we're a terrible public speaker across the board. And because they think we're a terrible public speaker, we assume they must also think we're terrible at everything else in life.

We're laughably wrong. The scientists found that most people aren't all that judgmental and quickly forget single errors. But because we overblow the implications of every social move we make—because we believe that everyone cares so deeply about our every public action—it causes us anxiety and stress.

The scientists call this the "overblown implications effect." It's a wing of the "spotlight effect," which is how we overestimate how much other people think of us. It's as if we believe we were living in our own prime-time television show—*The [Insert Your Name] Show*—with the spotlight always on us. But the reality is this: we're usually too blinded by our spotlight to stare at anyone else's.

There's also what psychologists call the "fundamental attribution error." It's how we attribute other people's actions to their character but attribute our actions to factors outside our control. When someone else is late to a meeting, they're lazy. When we're late, it's because of traffic.

Or there's the "overconfidence effect." This is how we tend to have excessive confidence in our beliefs. Scientists found that people who said they were "99 percent certain" on an answer were wrong 40 percent of the time.

Or there's the "false uniqueness bias," which is our tendency to see ourselves and our work as more unique than it is. It often leads us to focus on the differences we have with people rather than our similarities. Which explains the concept of schismogenesis, the idea that cultures and people define themselves against their neighbors.

Or there's "naive cynicism." This one's when we think everyone but us is selfish. That's even though many of us, as the writer David Foster Wallace put it, tend to be "operating on the automatic, unconscious belief that [we are] the center of the world, and that [our] immediate needs and feelings are what should determine the world's priorities."

We are all, it seems, plagued by "naive realism," the belief that we see reality as it is. Nope. We don't.

Psychologists have discovered hundreds of these cognitive biases. Start stacking them up, and we can see how it might be possible that, yeah, perhaps our drive for influence does lead us astray.

As I considered how my own drive for influence has led to highs and lows, I thought back to something a friend told me.

About five years ago, my wife and I had a dumb argument. I wasn't backing down. She wasn't backing down. It was as if we were both sipping strong cocktails of the fundamental attribution error, overconfidence effect, and naive cynicism.

During the stalemate, I vented to this friend. I explained to him in agonizing detail why I was right, why my wife was wrong, how the world would be better off if I could just get her to understand this—and did this guy have any advice for convincing her I was right?

His response: "Do you want to be right or happy?"

This question has since saved me a lot of headaches my ego-driven brain manufactures and seems intent on worsening by defending its position. And it highlights something important about our scarcity brain and its desire for influence.

Scientists at the University of Pennsylvania theorize that our capacity to reason didn't necessarily develop so that we could find better beliefs and make better decisions. Reason might have evolved so we could win arguments and gain influence. In our interactions with others, scarcity brain acts like our puff person, weaponizing reason to protect our status and make us look and feel good in the short term.

And this extends far beyond debates with our significant others. We're now out of our 150-person tribes where we'd all work toward a common goal. Our influence drive affects the sweeping range of interactions we have with others in person and online. Big arguments, yes. But more commonly the most minute differences.

We still default to one-upping to boost our influence. But to do so, our brains pull from all sorts of those cognitive biases. These mechanisms likely made sense during the crucible of human evolution. Back then we were far more likely to debate positions that really did affect our survival. Influence could mean life or death.

But in our safe, comfortable world, most of our everyday disagreements are astoundingly inconsequential. Yet we don't see this. We let our ancient drive to always look good and maintain influence walk us into anxiety, resentments, and misery. It hurts us in the long run.

When we ask ourselves, "Do I want to be right or happy?" we take the long view and insert perspective into the equation. But we can also bend the question. It could be "do I want to look good or be happy?" Or "do I want to one-up this person or be happy?" Or "do I want to be right or be a good friend, co-worker, or significant other?" And on and on. Play with it.

Choosing the latter option can be uncomfortable in the immediate short term (we're fighting against our pit bull brain). But over time, it has a way of dissolving the bullshit that causes our everyday suffering. And when bullshit dissolves, it becomes fertilizer, bringing growth.

"Do I want to be right or happy?" can even give us perspective and clarity to see another important fact: We probably *aren't* right in most arguments. And neither is the other side. Time changes our worldviews. Most of us can look back on past arguments and realize that there are very few where we were totally, undeniably, universally right. We overreact more than we underreact—another tendency that helped us survive in the past. Like a smoke alarm. It rings the same whether the smoke is from lightly burnt microwave popcorn or a massive blaze.

And who we are and what we know and hold true is a moving goalpost. A hill we'll die on today is one we'll happily cede tomorrow. Yet we suck at seeing this in the moment.

So "do I want to be right or happy?" is now a question I try to ask myself whenever my desire to influence others is pulling me in the wrong direction. My brain asks me to play one game, but I can quit it or force it to play another.

I am not perfect at asking this question. Wouldn't even say I'm good at it. But remembering to ask myself if I want to be right or happy when I'm in an ego trap, although uncomfortable in the moment, buys me some emotional space later. It lends perspective and cuts down on my daily suffering. And that feels like a win for us all.

As I was driving home from my meeting with Nguyen, I noticed that I still, quite embarrassingly, had croissant crumbs on my jacket. My mind drifted back to Iraq. To a bakery that my fixer took me to. Erbeed's approach seemed to be to feed me to buy time so he could frantically attempt to orchestrate meetings from nothing.

The bakery was near the Al-Kadhimiya Mosque, one of the holiest sites in Shia Islam. The smell of flour, butter, and sugar met us as we entered. Loudspeakers mounted to the mosque's minarets were calling all to prayer, reciting words from the Koran, a beautiful song that reverberated from the store's tiled floors.

Glass cases formed a long horseshoe around the shop. They were packed to the brink with "sweeeeets," said Erbeed. "Yessss, sweets."

There were at least a hundred baked goods. They were glazed, seeded, honey topped, frosted, filled with fruits or nuts. Some of them were neon yellow, cerulean blue, sea foam green, coral red, marigold orange. They were formed into circles, squares, rectangles, crescents, and the like. All of them were sold by the kilogram. The store owner began thrusting samples my way. "You eat," said Erbeed. "Yes, you eat. You Americans like the sweets and eat too much." He wasn't lying for once.

There's something else humans are built to crave. And if we believe the statistics, this scarcity loop is killing one of us every thirty-four seconds.

Food

Of everything we crave, food tops the list. We require a certain amount of food to survive. If we don't get enough, our health goes all to hell. The same goes if we get too much.

But for most of human existence, food was scarce. The trouble was always finding enough food. To improve our survival odds, our brains developed elegant machinery like the scarcity loop to help us persist in our searches.

Yet our food back then was a lot like those old casino slot machines no one played. *Finding* the food fell into the scarcity loop. But eating it? Not so much. Our food was boring. We were eating plants we found and rangy animals we hunted and fished. Think: Plain and dirty roots and some slimy fish cooked over an open fire. No salt. No sauces. Our modest diets weren't all that craveable.

Like slot machines, this changed starting in the 1970s. That's when the global food system tipped to producing an abundance of exceedingly delicious foods. Like me in that Iraqi bakery, people worldwide suddenly had hundreds, even thousands of options of exquisite mixtures of sugar, salt, and fat. We whipped our potatoes with butter and sour cream and topped them with cheese and bacon. We breaded and fried our fish and dipped it in fatty sauce. We didn't have to work to find it. We didn't even have the pause and effort of needing to prepare it beyond opening a box or bag and starting a microwave. We hardly even had to chew it.

This delectable, preformulated food was everywhere, in mass quantities. Eating food adopted elements of the scarcity loop. The food itself was engineered to be irresistible and quickly consumable so we'd eat more faster. In the past, we might have eaten one or two meals a day. But we began eating around the clock. Research from the Salk Institute found that the average person today eats three square meals and various snacks across a fifteen-hour window. This new food incited quick repetition, pushing us into too much when we'd had enough.

Each meal could also become a sort of unpredictable gamble of deliciousness. For most of history, we ate the same boring things every day. But now we have thousands of options, each promising a new blast of flavors we've never experienced. Research shows that people eat more food if they have more different tasting options. It's why we tend to over-eat at buffets.

Scientists haven't pinpointed exactly how our shift to an abundance of delicious, scarcity-loop-inducing foods has altered our health. We know they have a big effect. But how big? And what would our health look like if each meal wasn't an explosion of sugar, salt, and fat?

We're still shaky on the topic because studying it perfectly isn't feasible. Most of nutrition science is based on flawed surveys. And diet-related killers like heart disease and some cancers develop over decades, not months or years. A near-perfect study would require locking people in a lab and feeding them the same foods for life.

But then I discovered the work of Michael Gurven. It turns out that the idea of learning how our preindustrial foods changed our health isn't so far-fetched. Except the lab is a handful of outposts in the most remote reaches of the Bolivian jungle. And the study participants are members of a mysterious and ancient tribe called the Tsimane.

But . . . *lo siento*. Sorry. No refunds.

My only option was to travel by land. It would take twelve hours instead of thirty minutes. I felt every second of the drive. It was down spastic, single-lane, vertical dirt roads. We passed white crosses embedded into the mountainside as the car's side tires rumbled inches from a cliff. Evidence of the dead lay in heaps of rusted old car frames hundreds of feet below.

Next the plan was to run this canoe six hours deep into the Bolivian Amazon. Eventually, I was told, the village of the people I'd come to meet would rise from a foggy riverbank.

We were navigating into a cool breeze and up the Beni River, a major tributary of the Amazon. We'd just run through a massive gorge. Then the boat's driver, Augustino, veered the *peque-peque* left up a feeder river of the Beni, called the Quiquibey. The water lightened in color and the wind warmed.

Augustino is eighteen years old. Rarely speaks. His boat, like all *peque-peques*, is a long-tail-style boat. Their name comes from their small motors, which make a *peque-peque-peque-peque* noise as they navigate waters. These motors have a propeller shaft that is six feet long and can swivel up and down and side to side. Hence the name, "long tail." This range makes the boat easier to maneuver and allows Augustino to quickly and easily lift the propeller from the water to avoid floating debris.

Which is good, because "this river is dangerous. It's high, fast, and filled with sticks and trees and boulders." That's according to Alex, a member of the Uchupiamona tribe who was raised in a village eight hours up another river. Alex was standing at the bow of the *peque-peque*, positioned like an NBA defender and acting as Augustino's eyes into forthcoming waters.

The river contracted and expanded like lungs. It deepened and sped in contraction, funneling logs and rapids at us. It shallowed and slowed in expansion, making room for partially submerged boulders and trees, or becoming so low that we ran aground.

"¡Ten cuidado!" Alex yelled at us to watch our heads as we passed under a thick tree branch above the channel. Alex is essentially a slightly less militant Che Guevara. He's an indigenous political leader and staunch defender of native tribes who fights against encroachment by a shifty government and the mining, logging, and petroleum companies it would like to quietly sell this land to.

"¡Allí!" he yelled, signaling a submerged log to Augustino.

Augustino came at the log fast and hard, and just when I thought we were all going overboard—*PEEQQUUEE . . . PEQUE . . . peque-peque-peque*—he symphonically throttled the engine and made a sweeping move with its propeller that put us in the perfect position to maintain momentum and miss the hazard by inches. The kid's the Dale Earnhardt of these waters. The Intimidator. He could put this boat in a space that would fit a Q-tip.

The banks of the Quiquibey are high and made of red clay. Atop them is a mass of fractaled green forest. Seemingly no beginning. No end. No points of reference. White birds followed us upriver as we crashed through rapids. We saw capybaras on the banks and turtles sunning themselves on logs. A human or two fishing from the banks or a canoe every hour or so.

After six hours of Alex yelling out dangers, we spotted a *peque-peque* and a couple of handmade stand-up canoes lashed together on a bank. Augustino swept the motor starboard, and we rode our vessel up a sandy beach. We all jumped out and climbed a sheer red bank. The smell of lemon trees met us as we reached the flat, impenetrable jungle.

"Vamos," said Augustino, flicking his hand toward a nearly invisible trail that cut into all the green. After a few hundred yards we came to a clearing. And then there they were.

The Tsimane who greeted us were shoeless and wearing worn clothing you probably donated to a Salvation Army in 2005. One wore Hollister jeans and a "Life Is Good" T-shirt. Another was in Old Navy corduroys and a white T covered by a Gap cardigan. Yet another wore Levi's and a polyester soccer jersey. There were ten of them. Three adults. The rest teens and children.

Augustino quietly approached the oldest man in the group and said something in Tsimane. The man listened while staring at the jungle floor. No reaction.

Alex whispered to me. He told me the man's name was Leoncio. Leoncio was the de facto leader of this fifty-person Tsimane village of Corte. He was about five feet two and 120 pounds. Dark brown skin and lightly salted black hair. Never raised his eyes as Augustino talked and occasionally nodded toward me. And I couldn't blame the man.

Gangly light skins like me bring trouble. Demands. Problems. We have since 1616, when Spanish explorers first contacted Leoncio's ancient tribe and offered Christianity in exchange for gold, land, and anything else they could slap a monetary value upon.

But Augustino is Tsimane. And Alex has a good reputation in these parts. So he stepped in and started speaking to Leoncio in Spanish. I pick up words like *periodista, corazones, arterias,* and *salud.* Journalist. Hearts. Arteries. Health.

A blaze-orange butterfly floated between them as Leoncio finally responded. "Sí," he said. Then his eyes walked up the ground and finally met mine. A knowing look took his face. The gangly light skin he was looking at was a dying man.

Assuming I'm like most other Americans, my arteries are slowly clogging. Accumulating plaque and fat. Disgusting, smelly, waxy white gunk that's a lot like what oozes from a pimple.

Eventually, one of these arteries will build up too much of this plaque. This will cut off the blood and oxygen to my heart or brain and leave the organ—and, by default, me—to die.

At best, I might survive this heart attack or stroke with a massive fright and ensuing regimen of pills and procedures. A less decent scenario is that I weather it with paralysis on one side of my body. Or an inability to speak or see. Or a brain that can't remember or think. Or in a sort of vegetative state.

At worst, it'll drop me on the spot. Widow-maker-style. *Muerte.*

There's roughly a coin-flip chance that one of these scenarios is my fate. I am not unique among Western people.

We call these problems of the heart and veins cardiovascular disease. And cardiovascular disease is—by far—the thing most likely to kill me, you, and everyone we know. It now kills more people globally than our eight other top causes of death *combined.*

Medical science is great. It's helped people die less often from other maladies like cancer, lung disease, and car and work accidents.

But it has trouble keeping up with cardiovascular disease. The disease has been surging across the globe, spreading into new places and more people. Its rates dropped as people began smoking less and we invented the medication statins. But it's now cranking back up. Globally, it kills 50 percent more people today than in 1990. That figure outpaces population growth.

The numbers are savage. In the United States, cardiovascular disease kills one of us every thirty-four seconds. Scientists at Tufts discovered that only 7 percent of Americans have "optimal cardiometabolic health." Even the young are at increasing risk. About 30 percent of all heart attack patients today are between thirty-five and fifty-four years old. Now 40 percent of all people who die before turning seventy die of cardiovascular disease. Exactly 44.7 percent of American women above age twenty have some grade of cardiovascular disease. In the United States at the height of the pandemic, the malady killed 250 percent more people than did COVID-19.

So, yes, cardiovascular disease is what kills us. But we, apparently, would rather ignore our fate.

Consider that the world's largest media outlets like the *New York Times* and the *Guardian* run nearly tenfold more headlines about cancer and twenty-fold more headlines about murders and terrorism. Yet the average American is 50 percent and thirty-eight times more likely to die of cardiovascular disease than of cancer or by homicide, respectively.

We also fret about the wrong things. The average American is nineteen times more likely to worry about cancer than cardiovascular disease—going down anxiety-ridden rabbit holes with Dr. Google and her diagnosis that each of our physical anomalies is surely stage 4 cancer. This is like obsessing over a strange-looking mole on your trigger finger as you play a slow game of Russian roulette.

There is a possible solution in this jungle. And Leoncio and the other members of the Tsimane seem to have it. His arteries are most likely the opposite of mine: clean as new copper pipes.

That's according to Gurven's team of doctors who recently studied nearly a thousand Tsimane over age forty. It was a study of epic proportions. "We really wanted to delve into heart disease at a much deeper level than just taking some blood pressure readings," said Gurven. The scientists gathered tribe members, ran them downriver, and rented a bus to take them to a medical clinic in the region's capital to analyze their hearts with detailed CT scans.

"We'd made this general observation that the Tsimane didn't seem to die of heart attacks," said Gurven. But he had no clue just how healthy their hearts would be. The scans revealed that the Tsimane had the healthiest hearts ever recorded by science. Even the oldest among the tribe, people well past seventy years old, showed little evidence of heart plaque buildup. Tsimane hearts appeared, on average, twenty to thirty years younger than those of the average American.

Even wilder, the effects of their healthy hearts cascade upstream to their brains and downstream to their other major organs. One study found that aging Tsimane brains decrease in size 70 percent slower than the brains of Americans. This means the Tsimane don't seem to get dementia and Alzheimer's (the fifth-leading cause of death worldwide). Another study found that they don't get diabetes and kidney diseases (the ninth-leading cause of death). Even the cancers that kill us are rare.

Hell, the Tsimane even have what's called a slower epigenetic aging rate. That's a detailed measurement of how our cells, tissues, and organ systems are aging.

Meanwhile, the average American is more likely to be sick than not. Six out of every ten of us have a chronic disease. Four out of ten have two or more. The Tsimane: zero for ten. They suck at dying of the diseases that kill us.

They've found enough in a way that we once did, but lost.

After some formal introductions, Leoncio led me onto a tight dirt trail walled by jungle. Decaying brown leaves covered the ground. From it sprang grasses, plants, bushes, and trees of all sizes.

Leoncio is fifty, but he's built like Bruce Lee and moves with the same quickness and grace. The man was bobbing and weaving and never breaking elegant stride. I, meanwhile, was moving with the grace of Lenny Bruce. Tripping, bumbling, and trying to stay on the trail and follow close enough to hear his mumbling.

One tree's trunk rose vertically nine feet, then broke into giant, six-foot-long, two-foot-wide leaves that fanned out from the tree's top. Another's burst a foot from the ground then quickly forked out into long, leafy branches that rose ten feet. Football-shaped brown nuts hung from the base of those leaves. Towering above it all were trees whose trunks were ten feet in diameter and climbed two hundred feet to top out the canopy. Biologists say there are up to sixteen thousand tree species in this jungle.

To my suburban eye, it was all just a big wall of green distinguished only by height, width, and leaf shape. But as we walked, Leoncio was humoring me, acting as the most unenthused park ranger of all time. He pointed at plants big and small and quietly muttered everything he and his people eat from each. "Banana, plantain, chocolate, avocado, corn, nui, caururu, *achachairú*, pacai, potato, yucca, rice," he said.

Some of this stuff grows naturally. Others he's planted, embedding it within all that green.

The forest cleared into an opening, revealing a few open-air structures. They were made of dark hardwood planks, and their roofs were thatched and floors dirt. The structures looked simple. But they were the pinnacle of craftsmanship. Alex mentioned that Tsimane thatched roofs can last thirty years.

Wiry chickens ran underfoot. Ducks waddled in frantic formation. A catfish, maybe fifteen pounds and recently pulled from the river, hung on a plank.

A fire pit sat in the middle of it all. A couple of Leoncio's children huddled around a big black pot on the fire.

They wanted to welcome me with a meal. But first we had to prepare it. And it was a group effort.

One of Leoncio's sons, who was twelve years old, was gripping a ten-pound wooden pestle shaped like a barbell and using his entire body to repeatedly thrust its rounded end into a big mortar filled with what looked like beige seeds. "Peeling rice," Alex told me. The family grows this rice just a hundred yards from here.

They asked me if I'd like to try. "Sí," I said and began thrusting. It was a workout that produced something tangible. The muscles in my back and arms tired after a hundred thrusts.

Then one of Leoncio's daughters, maybe ten years old, was gripping a machete and determining which of those small chickens would make a good lunch. She snatched one with a rapid death clutch.

Then Leoncio's wife threw that rice into the pot on the fire, and another daughter was set off into the jungle to pick some plants.

An hour later, we were all sitting down to eat. And even though we were in the middle of the jungle, this meal's basic makeup and flavors would be familiar to much of the world up to World War II.

Our lunch was various parts of that unfortunate chicken. I had its leg, and there wasn't much meat on the bone. The chicken was more like one we in the United States would have raised and eaten seventy to eighty years ago. In the 1950s the average chicken took roughly two months to reach its maximum weight of two and a half pounds. Today, breeders can use less food and bulk up a chicken to nine pounds in the same timeline.

A pile of rice and a pinch of onion and tomato flanked the meat. A communal bowl held plantains baked in their skin in the coals of the fire.

Before I arrived in Bolivia, I also spoke to Hillard Kaplan. He was Gurven's PhD adviser and also studies the Tsimane. His takeaway message: "Cardiovascular disease is a modern phenomenon and your lifestyle makes the difference in your likelihood of dying from it. And what's good for your heart seems good for the rest of your body and preventing other diseases."

It all starts with what we eat.

The thing I had trouble with, and why I took this most inconvenient trip down into the jungle, was this: What Gurven and Kaplan told me about Tsimane eating didn't add up.

Name a diet, any popular diet of the last fifty years. The Tsimane way of eating, somewhere in a day's worth of meals, will give it the middle finger. It's not paleo, vegan, keto, plant based, low carb, Mediterranean-style, or any of the fads we were sold and told are the key to health, longevity, and a body you could put on a magazine cover.

What Leoncio and the twenty thousand other Tsimane scattered throughout the Amazon eat even upends the nutritional research and advice of some of our most venerated academic institutions. Even the American Heart Association would think twice about giving it a full stamp of approval. And yet ...

As I gnawed on the leg of the chicken who was alive just sixty minutes earlier (RIP), I asked Leoncio what he eats every day. I wanted the rundown. "Dame el menu." *Give me the menu,* I said in my gringo-accented Spanish. Alex laughed.

Then Leoncio was giving me a grocery list. "Fish. Meat," he said. The man uses a traditional recurve bow to fish and hunt animals ranging in size from five to six hundred pounds. The largest animal he hunts is tapir, nicknamed the Amazonian deer.

The Tsimane eat fish or meat with breakfast, lunch, and dinner. Meat, of course, offends all vegan and plant-based diets. But it's also been inciting vitriolic debates among nutrition scientists since the 1950s. Today, scientists are engaged in what they call the Meat Wars over whether eating meat is healthy or a one-way ticket to disease and death. For example, researchers at Harvard and Texas A&M recently got in an ugly public spat over the healthfulness of meat.

Then Leoncio was onto fruits and vegetables, muttering names I knew and obscure others that will likely become the next "superfood" marketed to gullible Americans like myself. "*Majillo*, avocado, chia, grapefruit, onion," he went on. With a heart like that, I assumed he must eat many of them. So I asked, "Mucho vegetables?"

Leoncio just shrugged. "Eh, poco." *A little*, he said. In a full day he eats about as many vegetables as we might in a single salad.

Then Leoncio seemingly went off the rails. "Sugar. Chocolate," he said. He grows the cane himself. Harvests it, juices it, then dries that juice into sugar and uses it to sweeten juices and foods. Same goes for chocolate. One popular doctor claimed that sugar is eight times more addictive than cocaine and will kill you quicker. Leoncio did not appear strung out or clinging to life.

"At the same time, they aren't eating *a lot* of added sugar," Gurven told me. We, too, would probably all eat much less sugar if we had to personally grow, harvest, and process every gram we ate.

But all those foods are just a fraction of the Tsimane diet, Leoncio said. Maybe a third of it. He then started talking about the bedrock foods that keep him and his family alive and well. The ones he eats most.

And if there's one thing modern internet diet gurus seem to agree on, it's that these foods are no good, very bad, terribly awful. They're the types of basic foods that would send paleo diehards, keto zealots, "clean" eaters, and Gwyneth Paltrow and all her Goop followers into anaphylactic shock.

"Corn," he said. Eats it plain and makes a thick, semi-alcoholic drink called chicha out of it. Corn has become a four-letter word in food-conscious circles. It was off-limits in nine out of ten recent best-selling health books. Netflix as of this writing has various documentaries on how corn is, these films argue, a sort of creeping specter that's silently killing us all.

"Rice." But not the special kind you find at Whole Foods with all sorts of qualifiers like brown, wild, organic, non-GMO, or something called "forbidden." We're talking white rice. His family peels it. Even nutrition scientists at Harvard say we should all avoid white rice. They cite a largely unproven theory about weight gain called "the carbohydrate-insulin model of obesity." It states that carbohydrates elevate our blood sugar levels, which increases a hormone called insulin that makes us store fat. Leoncio eats rice once or twice a day and is as lean as a professional marathoner.

"Potatoes. Plantains," he said. These foods, boiled or baked and served plain, also end up on Tsimane plates once or twice a day. Potatoes and plantains are both, as the Harvard scientists put it, "starch bombs." Starch is a carbohydrate in everything from wheat to rice to corn. But foods like potatoes and plantains have relatively more of it. Some scientists say starch is "bad" based on that same controversial carbohydrate-insulin model theory. They also refer to studies that found that people who eat a lot of potato chips and French fries tend to gain more weight over time. Which . . . if we banned every natural food that we also happen to fry, coat in cheese, or mix with cream, butter, or sugar, we'd have nothing left to eat.

Gurven told me that, on balance, about 70 percent of the Tsimane diet comprises these carb-filled, boring foods. Foods that modern diets and in some cases even our most respected institutions have vilified. And, I'd learn, quite wrongly.

To understand why Tsimane hearts don't keel, we have to understand scarcity brain on food. How our strange new world of abundance of calories and tastes and textures is altering systems that developed over half a billion years. This can also reveal why the diet and weight loss industry, worth $800 billion a year, has mostly been a spectacular failure. We are, in fact, projected to rack up more diet-related diseases over the next few decades.

Scarcity brain evolved in a world where food was often in short supply. So it's built to crave all foods, but especially foods that are packed with calories. The more rich and calorie dense a food, the more delicious it is. The more delicious it is, the more likely we are to crave it and, in turn, eat a little more than we need. Our body stores that extra food as fat. In the past, having extra fat was an insurance policy against starvation. Our bodies would draw on that fat to survive food scarcity. For energy when we couldn't find food.

But through basically all of time, the food we hunted, pulled, or picked from the earth wasn't that rich and calorie dense. It wasn't just scarce in quantity. It was also scarce in calories, tastes, textures, and all the other qualities that make food more enjoyable. We were eating plants we found and animals we hunted and fished. Remember: dirty roots and some slimy fish cooked over an open fire. Our modest diets weren't all that craveable.

At least thirteen thousand years ago, we began realizing that we might be able to have more food if we raised and grew some of it rather than just hunting and stumbling upon it. So, like Leoncio, we continued hunting, fishing, and gathering. But we also started raising animals like sheep and goats and growing foods like rice, potatoes, corn, and wheat.

Humans began farming wheat, peas, lentils, and chickpeas in the Middle East. Potatoes, beans, corn, and peanuts in Central and South America. Rice in China. And so on. These farmed foods led to population growth and human civilizations and empires. The food sources quickly spread around the world, thanks to trade and conquest.

Humanity began leaning into our farmed foods because they are amazing plants, superfoods in the sense that it is super to make progress, live well, and not die. "They're the only way you can have cities and armies, and those are what create empires," Rachel Laudan, a food historian, told me. Agriculture scientists call these foods staple crops. They became staples of the human diet because they can deliver many people enough calories, vitamins, and minerals to thrive.

It's now popular among some academics to say that farming was the "worst mistake" humans ever made. They argue that it led to problems like class stratification and that our hunter-gatherer ancestors worked less than us. But there's no strong evidence that our ancestors worked less than us (most studies suggesting this don't consider how long it takes to process the hunted and gathered food). And back before we started farming, scientists estimate the world had about ten million people. This is because hunting and gathering typically require massive swaths of land to find enough food to not starve. Had we not started growing staple crops, we'd still be out there digging up and chasing around our food. And there'd still be about ten million of us. Chances are you, those academics, and I wouldn't have been one of those lucky ten million. So let's not complain too much.

Vegetables and meat were generally small side dishes. Vegetables have important nutrients but don't give us enough calories for the effort. For example, wheat has five to twenty times more calories per serving by weight than most vegetables.

Which is why humans until just recently would starve if they tried eating like an L.A. vegan, with only colorful plants all the time. That's only possible if you have a well-funded credit card and easy access to modern grocery stores and yuppie fast-casual salad restaurants. So the Tsimane eat vegetables the way humans have for aeons—as a supplement that gives them a bit of a nutritional boost.

Same goes for meat and fish. Despite what keto, paleo, and carnivore advocates believe, most early humans weren't eating big servings of animal flesh. They weren't walking around gnawing on cooked legs of the animals they'd hunted, as we often see in cartoon depictions of our cavemen ancestors.

Alyssa Crittenden, an anthropologist at UNLV and expert on the evolution of the human diet, told me that those popular diets focus far too much on meat. "Hunter-gatherers generally eat mostly plant-based diets," she said. Foods like tubers and fruits. There are rare exceptions, like tribes in the arctic regions. But even in rare times when meat is most abundant, the diet of most hunter-gatherer tribes tips more heavily toward plants. For example, even when meat is abundant for the Hadza tribe of northern Tanzania, it makes up just 40 percent of their diet. And that's only for the men. The women eat much less. Other times of the year the tribe relies almost entirely on wild tubers.

Meat and fish have necessary protein and minerals. They've been critical to the human diet. But the thing about animals is that they can run and swim away. Hunting is challenging even when you have a rifle, much less a homemade weapon like a bow or spear. Even once humans domesticated animals like pigs and cows, we couldn't raise that many of them. We didn't have modern bioengineering and an abundance of feed.

"Meat has always been very scarce," said Laudan, the food historian. "Much of the world was essentially vegetarian. You didn't want to kill off the animals you had, because they could give you eggs or offspring." Until 1900, she said, meat was generally considered a treat. It still is in many low-income countries. Eating as much meat as the average American does today—twelve ounces a day—was only possible for kings and queens.

In our past, most of us had diets like the Tsimane: staple foods like grains or potatoes supplemented with a bit of meat and vegetables. But these diets allowed civilizations to rise and fueled our greatest thinkers. Homer, the ancient Greek author of the *Iliad* and the *Odyssey,* called a diet of barley meal and wheat flour "the marrow of men." Even our fiercest warriors and conquerors ran on our staple crops. Roman soldiers ate two pounds of wheat per day. The British Empire, in the eighteenth century, took over a quarter of the world fueled by wheat bread. The Japanese samurai lived on rice.

Our staple foods even built our amazing brains. A recent study by a team of archaeologists at Harvard found that between 2 million and 700,000 years ago, humans began eating primarily starchy carbs like roots, similar to today's sweet potatoes. This may be what led our brains to double in size rather than the fatty meat some claim is entirely responsible for our brain growth.

Our environments of scarcity and something about our staple foods kept our hearts ticking. As a study in the *American Journal of Medicine* put it, "Heart disease was an uncommon cause of death in the US at the beginning of the 20th century." The disease didn't even show up in medical textbooks until 1904. But something changed in the 1970s.

Lunch was over. Leoncio and I were relaxing and watching his dogs vie for our leftover scraps.

I was full but not stuffed. Which made sense. A now-famous Australian study suggests that plain staple foods paired with some meat and vegetables are perhaps the most potent combo to fill us up yet keep us from overeating.

But I was also becoming skeptical. I wondered if the meal I just ate was extraordinary. Special. Something grand the Tsimane served to a guest but didn't regularly eat themselves. Perhaps Leoncio and his family simply didn't have as much food as us. We know that the less a person eats, the less likely they are to get heart disease. For example, during World War II, Norway saw its cardiovascular disease deaths plummet after it rationed food.

I asked Leoncio as much. In response, he swept his hand toward the jungle. One tree contained more plantains than you'd find in a bin at Costco. Another bore enough avocados to supply every Brooklyn hipster with avocado toast for a month. The rice was stacked deep.

Until the 1940s, most Americans didn't worry whether food would blow up their waistlines or clog their arteries, Adrienne Rose Bitar, a food historian at Cornell University, told me. People before then were far more likely to die in a sort of gross, theatrical fashion. They'd start vomiting and catch a fever or diarrhea. Or their gums would start bleeding out. Or they'd begin coughing up blood or developing terrible sores all over their bodies. These were the signs of cholera, dysentery, pneumonia, pellagra, scurvy, and so on. Most people at the time died of infectious diseases and malnutrition.

Food was considered medicine in the sense that eating more of it moved you away from death or helped you recover from our nutrient deficiencies and microbial killers. This is why the government suggested we eat more, giving advice like "eat a lunch that packs a punch!"

Yet a growing number of Americans and people in some other developed countries were dying strangely. There was no dramatic run-up to their death, where substances spewed from every orifice. Or where they looked and felt like the walking dead. These people would seem fine one moment and the next clutch at their chest and fall dead on the spot.

Just under 1 percent of people at the beginning of the twentieth century died of what doctors were calling diseases of the circulatory organs. But the number kept creeping higher. Food had something to do with this.

Humans have been processing food for 1.8 million years. Cooking, drying, grinding, fermenting, and more. But this processing was uncomplicated and done at a relatively small scale. Once the Industrial Revolution happened, we started making more of everything in different ways and in much larger quantities. Including food.

Wheat, potatoes, corn, rice, and oats were grown at a grand scale, then fabricated into thousands of foodstuffs. The average grocery store today has fifty thousand items. Most of them are delicious, packaged concoctions made from those five staple crops.

We altered our animals and the products we got from them. Meat in the ancient past wasn't like today's meat. We were eating rangy animals that—much like us humans in the past—moved around all day looking for food that was scarce. Our animals weren't overengineered, overmedicated, and overfed for delicious fatness. For example, due to the extra fat, modern cuts of steak can have about 87 percent more calories than game meat like that the Tsimane eat.

On factory floors, at home, and in restaurants, we began morphing the staple foods that sustained healthy human life for thousands of years into complex dishes and packaged foods that were unknown to scarcity brain. Foods our scarcity brains are built to crave. Calorie-packed concoctions of sweet, salty, fatty, crunchy, tangy, and so on.

We tend to hear all negatives about our modern food system. But the upsides of this food revolution were massive. For most of the time before this, "our ancestors lived mean, short lives, constantly afflicted with diseases, many of which can be directly attributed to what they did and did not eat," Rachel Laudan, the food historian, wrote in a famous essay.

Food became abundant and hunger began dropping around the globe. It became safer and easier to transport and lasted longer. It became more nutritious. We learned that strange diseases like goiter, scurvy, and pellagra are caused by a deficiency of single nutrients, like iodine or vitamin C or B_3. We began to fortify foods with the nutrients we lacked and—poof—we saved millions of lives. People even started getting taller.

World War II amplified food processing. It was a massive exercise in feeding millions of troops around the world. Small food corporations joined the war effort to help supply troops with long-lasting, transportable rations. They were General Mills, Kraft, Nestlé, Coca-Cola, and so on.

After the war, with massive new production lines, these corporations turned their attention to feeding the average American. They leaned on advertising agencies to pitch their products. Food and agricultural scientists jumped on board (by the end of the century, we had seventy thousand of them in the United States alone). An industry bloomed.

By the 1950s, the modern food and medical systems were rendering malnutrition and infectious diseases moot. We had plenty of food and ate it as if times were still lean. And that's about when Ancel Keys, who had taught at Harvard and was a physiologist at the University of Minnesota, noticed a paradox. Wealthy American men with ample access to our new foods were well fed. But they were ravaged by those clutch-at-the-chest, fall-dead-on-the-spot deaths.

Poor men in small towns in southern Italy and Japan and other countries, on the other hand, had anywhere from four to ten times less risk of cardiovascular disease and often lived past ninety.

The success of fortification had shown scientists that not having enough of a single nutrient within food could cause disease. But, Keys wondered, could having *too much* of a nutrient do the same?

Keys thought so. Fat made up nearly half the calories in the average American diet but just around a fifth in the Italian one. Heart disease, wrote Keys, "is greatest among American men and tends to become progressively less as we pass to countries where fats contribute less and less to the total diet."

"Then, in 1955, President Eisenhower had a heart attack," said Adrienne Rose Bitar, the Cornell food historian. "That's when public attention cohered around the idea that heart disease was an epidemic. Eisenhower got put on a low-fat, low-cholesterol diet. Dr. Keys's work got picked up in popular culture and began shaping government nutrition advice."

The scientific thinking around food shifted. We started seeing food as a bunch of individual chemicals we could formulate to make us healthy—or unhealthy. And so the government, universities, and diet gurus swooped in to deliver advice to eat less or more of specific unseeable components within our food. Like fat, carbohydrates, saturated fat, fiber, sugar, and protein. From this came all sorts of diets sold to us claiming to hold the secret to health: eat low fat, or low carb, or more protein, or less sugar, and on and on. It was like Nguyen's ideas around value capture, but with components of food.

By the 1970s, something had fundamentally altered our ability to find enough. "If you look at the pattern of obesity, we've been getting fatter for the better part of a hundred years," said Stephan Guyenet, an obesity researcher and neuroscientist and author of *The Hungry Brain.* "But obesity accelerated in the '70s, '80s, '90s, and through today. What caused us to start eating more? I think there are various explanations, but it clearly has something to do with our food."

In the late 1990s, Kevin Hall was finishing his PhD. Cardiovascular disease had dropped from a high in the 1950s because people were smoking less. But over the last couple decades it had been picking up as obesity grew.

Hall is a physicist by training, a hard-core numbers, data, and figures guy. He's a researcher at the National Institutes of Health. This role affords him big funds and resources to do the kinds of tightly controlled experiments that can dive into the nitty-gritty of food's effect on our health and waistlines. He locks people in labs. But, for ethical reasons, he can keep them there for only so long.

When I spoke with Hall, he told me, "Typically, nutrition science focuses on trying to understand the molecules in the food—the carbs, fats, protein, fiber, sugar, sodium saturated fat, et cetera, et cetera, et cetera—and how the body processes those individual nutrients and whether higher or lower values of any of them is healthier." Hall had famously discovered, for example, that there's no difference in weight loss between low-carb and low-fat diets so long as a person eats the same amount of calories.

But in the early 2010s, a group of scientists from Brazil began arguing that, as Hall told me, "we were thinking about it all wrong. The group said that the individual nutrients within food aren't important or even interesting. They said that the healthfulness of food is really about the extent and purpose of its processing. They said that ultra-processed foods are what's driving obesity, driving cardiovascular disease."

"Ultra-processed foods" are officially described as "formulations mostly of cheap industrial sources of dietary energy and nutrients plus additives, using a series of processes and containing minimal whole foods." Which is to say that the term is a scientific euphemism for "junk food."

For Hall, the theory from the Brazilian scientists screamed of more bullshit ideas being piled onto the mountain of them in nutrition science. The laws of thermodynamics say that a calorie is a calorie whether it comes from fresh broccoli or a Hostess brownie. "So when I first heard their claim," he said, "I thought it was nonsense."

He even called the group to air his grievances and ask hard questions. He pointed out that humans have been processing foods for a long time, yet obesity rose only recently. "I asked them, 'What is it about ultra-processed foods that you think is bad?' And the answers I got all seemed to undermine their argument. They would say, 'Well, these foods are really high in sugar and salt and fat and they're low in fiber and protein.' And I would say, 'Well, wait a second, you just named a bunch of nutrients. You can't say it's not about the nutrients, and then, when I push you, the first thing you say is that these foods are all high in these nutrients that are bad and low in nutrients that are good.' It struck me as intellectually lazy."

The only scientific way to shut up another person, especially if that person is a scientist, is to prove them wrong with data. Hard numbers and figures.

"So we brought twenty people into our lab, and they stayed with us for a month," said Hall. "And we gave them three meals and a box of snacks every day and simple instructions to just eat as much or as little as they wanted." Tightly controlled research like this is more reliable than standard survey-based nutrition research. That's because scientists measure exact quantities of food and control for all of the other factors that could influence weight, like exercise, smoking, or stress.

For half the month, the participants' meals and snacks were a standard American diet filled with ultra-processed foods. For the other half, they were what the Tsimane might eat.

Breakfast for the standard American diet weeks was something like croissants with butter, fatty pork sausage, and bright blue Yoplait yogurt. During the Tsimane weeks it was eggs with potatoes and vegetables. A junk lunch one day was a cheeseburger with fries. A Tsimane lunch was salmon with a sweet potato and green beans. One junk dinner was a fatty burger with buttery mashed potatoes and gravy. A Tsimane take on that same dinner was a leaner cut of steak paired with plain rice and vegetables.

"The two diets were matched for the amount of calories, carbs, fat, sugar, sodium, protein, and fiber," said Hall. "So I thought, if it's really these nutrients that are driving weight gain, then there wouldn't be any difference."

Hall and his team tracked, weighed, and logged every morsel of food the participants ate.

And, yes, the standard American diet had cookies for dessert—and graham crackers, candied fruits, or peanut butter snacks. Dessert for the Tsimane group, on the other hand, was plain yogurt with some berries, or orange or apple slices.

The results? If you give a human a cookie—or cheeseburger or royal blue yogurt or mashed potatoes injected with butter and cream and topped with thick salty gravy—we will eat more and more of those foods until we fatten up and die of heart disease. If you give a human plain yogurt with some berries—or plain potatoes, lean meat, or rice—we will eat just *enough* of those foods. We'll be less likely to fall into a scarcity loop of food.

The people in the study didn't so much eat the junk diet as pound it. "The people on the ultra-processed diet ate five hundred calories more per day and they gained weight and body fat," said Hall. But when they ate a Tsimane-like diet, "they spontaneously ate less and lost weight and body fat."

The possible explanations all seem to go back to scarcity brain and our world of abundance.

One explanation, said Hall, is that ultra-processed foods have an abundance of calories in every single bite. "When you create ultra-processed foods, you concentrate the calories," he said. "So every bite of food, no matter the nutrients, has more calories." For example, let's say we eat two ounces of potatoes. If they're in plain baked form, it would be 50 calories. In potato chip form, it's 360 calories.

The second possible explanation echoes the quick repetition embedded in the scarcity loop. "We found that people ate the ultra-processed meals a lot quicker," said Hall. This could be because the Tsimane foods led the participants' brains to pump out more of a hormone called PYY, which reduced their appetite. They also decreased a hormone called ghrelin, which made them hungry. The Tsimane foods even took more work to physically chew. The ultra-processed diet, meanwhile, did the opposite. It cut those natural brakes that help us find enough.

So what accounts for the rise in obesity and its heart-stopping side effects? By the 1970s, our scarcity brain—which for millions of years tuned itself to a world of less food and less interesting food—was living in a world of abundance. Food engineered to lead us to eat more faster was everywhere and readily available to most Americans. And we ate it more often.

Quick repeatability was key for the food industry. Take snacking. The food R&D executive Carlos Barroso said, "There are three Vs in the snacking world: value, variety, and velocity." The scarcity loop.

"Beginning in the 1970s, you see really profound changes in snacking," said Guyenet, the nutrition researcher. "There was this huge marketing push to create an entirely new food category in the American diet, snacking. It was successful. And if you look at the types of things people snack on, it's ultra-processed foods."

A calorie is still a calorie, said Hall. Junk foods don't break the laws of physics. But we'll eat more, faster if a calorie comes from junk. Hall is now designing studies to figure out exactly what is driving the weight gain he observed in the study.

Yet still, when we want to reclaim our health, we're sold some fad diet that demonizes or worships one nutrient or aspect of food and provides us with all sorts of ultra-processed foodstuffs that fit our diet's regulations. For example, Burger King will sell you a single keto-diet-style sandwich that contains more calories than the Tsimane have in two full meals. Or stacks of paleo pancakes. Followed by keto candy bars. Then low-carb tacos. Then low-fat potato chips and cookies. Bars like those made by Pure Protein have the same amount of calories and processing as candy bars, and they're the second most popular snack on Amazon.

And this is, as the scholar Yuval Noah Harari explained, a double victory for consumerism. "Instead of eating little, which will lead to economic contraction," he wrote, "people eat too much and then buy diet products—contributing to economic growth twice over." When we buy "diet foods," we often pay more to eat less, a strange phenomenon in the grand scheme of things.

And when our diets get too extreme, when they mindlessly force us into less, our scarcity brain goes on the defensive. We learned this from the Minnesota Starvation Experiment, conducted in World War II. It found that when we eat too little, our body slows its metabolism by turning down health-promoting bodily processes. It does everything from lowering our heart rate and body temperature to shrinking our organs. Meanwhile, our brain responds by dialing its attention to food and increasing our hunger signaling, making food harder to resist. The result: weight loss slows, hunger grows, and we usually fail and rebound hard. The study participants ended up heavier than when they started the study. And their metabolism and hunger signaling was off for years.

The great irony is that those effects of starvation are beneficial if a person is *actually* starving. They helped us save energy and prioritize food back when food was harder to come by (long before we found ourselves in a world of stocked pantries and fast food on every corner).

Today, however, these scarcity brain survival mechanisms are working against us. They're why so many of the 74 percent of Americans who are overweight or obese struggle to lose weight. Dieters walk blindly into "less" and start with enthusiasm and a normal metabolism. Then they start to drop weight fast and their body sabotages their efforts.

The scarcity loop helps explain why 95 percent of people who lose weight in a given year eventually gain it back. When we begin to lose weight, the scale gives us unpredictable rewards every morning. Our weight changes unpredictably, trending downward in an exciting way. But when weight loss stagnates, the number becomes predictable. The unpredictable rewards no longer trickle in. Dieters get frustrated—and extinction occurs. They slide back into their old habits. A solution: Stick to your diet but find *another* behavior that rides the loop and aids your goal. For example, lifting weights. As you become stronger, the amount of weight you can lift and number of times you can lift it changes unpredictably and trends upward in an exciting way.

I spoke with my friend Mike Roussell about our new food and diet environment. He's a PhD nutritionist who runs a nutrition consulting business and works with average people, professional athletes, and Silicon Valley–billionaire types who make him sign nondisclosure agreements.

"My dad has been a tailor for forty-five years, and he's been busier than ever since the pandemic," said Roussell. "No one fit into their clothes. The combination of inactivity, being around ultra-processed food all the time, and using food to relieve stress has led tons of people to gain weight. But the worry is that those new habits last."

For example, candy sales hit a record during the pandemic. They boomed 15 percent. The candy trade group the National Confectioners Association wrote, "When times are tough, people turn to sweets to make themselves feel better. The pandemic not only gave people more license to buy goodies, it got them into the habit of buying them online and consuming them at home."

They're correct. Imagine the following scenario: We're sitting at a counter. In front of us are bowls of two new kinds of M&M's. The M&M's in the first bowl are "low calorie," while the ones in the second are "high calorie." Scientists at the University of Miami found that when we receive cues of scarcity—say, a news story that mentions that all is not right with the world, such as economic decline, a pandemic, or really anything about American politics—we choose the high-calorie M&M's and eat double the amount we would if we didn't receive a scarcity cue.

Negative information, the scientists believe, triggers our brain to unconsciously assume a famine is nigh. This reaction to scarcity cues was a feature that first came about long before humans. Pick any animal. When it thinks resources are scarce, it responds by eating. A lot. It attempts to put on weight. Bulking up is a brilliant defense mechanism.

Pair our scarcity brain with the modern news cycle, the rat race of life, abundant ultra-processed food, and the limited-time release of the McRib. Congrats, you have an elegant formula for folks who waddle.

Roussell recently challenged his clients to go a month where 80 percent of what they ate were Tsimane foods.

"One of the biggest things to come out of it was that people realized that so many of the foods they thought were 'healthy' were ultra-processed," said Roussell. "I had a guy on the challenge who ate a ketogenic diet. He discovered that most of his food was very processed because he was eating keto cookies, keto drinks, keto bars, keto packaged meals. Or there was a lady who would leave the office and go crush a workout at the gym. Then she'd buy this smoothie from the gym and go home and have a meal. The smoothie advertised itself as healthy and touted all these different health benefits, but it was full of added sugars and processed fruit goos, which made the smoothie check out around an unexpected nine hundred calories."

Roussell continued: "People struggle to see through all the health marketing. But, usually, packaged 'health' foods are actually more processed, because it's way harder to make a cookie taste decent if you're not using flour, sugar, or dairy, or any of these things. So I think the brain is getting impacted on so many levels: from a marketing perspective, but also because so many of our 'health' foods are hyper processed and therefore so easy to overconsume." For example, research from Cornell University shows people falsely believe organic processed foods have fewer calories and because of this eat more of them.

Roussell's experiment was like Hall's, but in the real world. "I didn't tell anyone to eat any more or less food," said Roussell. "But people found that they started losing weight without even paying attention because they were consuming fewer calories."

And unlike those Minnesota Starvation Experiment dieters, by eating a Tsimane-like diet, we don't lapse into the pernicious mind-body effects of eating too little, like we would on a crash diet. We find *enough.*

When I asked Roussell for his theories around the Tsimane, he answered quickly. "It's not rocket science to figure out why we overeat and they don't," he said. "Ultra-processed foods are fine-tuned at multiple levels to make us eat more of them. Their flavoring systems are dialed into what lights up our brains and makes us want to eat more. The foods are also physically processed in a way that requires less chewing, making it literally easier for us to eat more, faster. The Tsimane don't have these ultra-processed foods that are high in calories or easily accessible."

But over time, our brains have adapted to our delicious junk foods. "They push us to eat more," said Guyenet. "And our brain gets accustomed to these foods. Once you've been eating super-processed foods, you're not going to want to eat plain brown rice, broccoli, and fish. It's similar to what happens with addiction."

Once we know this, it's easy to get frustrated with the modern food system. We shouldn't. "The biggest misconception I see about food is that people think we have terrible food and an absolutely terrible food system," said Rachel Laudan, the food historian. "We have better food than anyone else in history ever." We just don't realize it, because few of us have experienced prolonged and serious food scarcity. Or read reports from the nineteenth-century coroners who were trying to figure out why so many young corpses were covered in the horrible sores caused by pellagra.

Abundance isn't just an issue in rich countries, said Laudan. Globally, obesity has tripled since 1975. Today more than half of all nations have obesity rates over 20 percent. The world now has 400 percent more people suffering from the ill effects of overnutrition, diseases of abundance, compared with malnutrition, diseases of scarcity. Food scarcity is mainly a problem of distribution and politics rather than availability. In the United States, we have so much food we throw out about a third of it.

Our problems of food abundance are good to have. But problems nonetheless.

Food corporations created and marketed hyperstimulating food. But, like gambling and drug use, blaming the industry entirely is just a way to off-load personal responsibility (that's one thing scarcity brain wants less of).

For example, it's common to vilify, say, McDonald's or Lay's for spending millions of dollars researching and testing fries and chips that offer the perfect combination of sugar, salt, fat, and texture so we reach, as the food psychophysicist Howard Moskowitz put it, the "bliss point." But do we really want to live in a world where we *don't* have foods that offer the bliss point?

"I don't have any gripes with the food industry making delicious food. That's their job," said Roussell. "But then it's our choice to then say, 'I choose to not eat that food every day and just enjoy it occasionally.'"

"And if you look at the 'foodie' industry of cookbooks, magazines, restaurants, chefs, YouTube, and Instagram," said Laudan, "you'll see this foodie community puts as much emphasis on taste as big food corporations." She points out that cookbooks for general readership, i.e., not rich people, until about (you guessed it) 1970 rarely made the way the food tasted the main consideration. "Saving money came first and taste was something that you hoped to get while being thrifty, but this has now flipped." That's another benefit of abundance: Food became cheaper. In 1920, Americans spent more than 40 percent of their income on food. By 2020, the figure was 8.6 percent.

So it's not as if we all decided, "Today I'm going to eat more!" Rather, we became able to buy and eat more foods that have an abundance of triggers that make scarcity brain happy and push us into an eating scarcity loop.

And why wouldn't we overeat those foods? Eating heavily when given the opportunity kept humans alive and well for millions of years. But for most of us, it outpaced our brain's innate ability to find enough. That extra food went into our waistlines and arteries. When you offer humans more, richer food, we will eat more, richer food. Unless we become aware of scarcity brain and elements of the scarcity loop baked into food—how scarcity pushes us to crave foods that only set off a spiral of more craving—and choose to eat like Leoncio. To eat enough.

We'd just finished lunch. "Okay," I told Leoncio. "I'm sold."

Now down to specifics. "¿Qué comes todos los días?" *What do you eat every day?* What does a day of Tsimane eating look like? I needed the details. He obliged and told me . . . nothing special.

For each meal, a quarter of his plate is usually a lean protein, like chicken, fish, a couple eggs, or red meat. Another quarter of the plate is vegetables, like cabbage. Holding down the final half of his plate are unprocessed staple crops, like potatoes, sweet potatoes, rice, quinoa, squash, corn, or plantains.

This diet is not as fantastically flavorful as we might be used to. But it's worked for thousands and thousands of years to keep us disease-free. And we need not obsess over the specifics. The only commonality between all of the Tsimane foods: they have just one ingredient.

I spent a handful more nights in the jungle. I ate more bland but satisfying meals. I accidentally stepped into an ant colony. The ants, in turn, savaged my leg with hundreds of bites. I watched a thirteen-year-old Tsimane boy use a machete to violently hack the head off a fifty-pound bagre catfish he caught. The sun was setting over the river as he hacked and hacked, sending up an eruption of blood and guts and fragments of spinal cord. It was all quite idyllic.

Although the Tsimane don't get the diseases that kill us, they of course still die. Gurven told me that pneumonia and other infections are still a problem. Or they face accidents. When I asked Alex about the dangers the Tsimane face, he turned and looked at me gravely. "Bushmaster," he said. It is, apparently, a pit viper that can grow up to fifteen feet long.

"Can you die from a bite?" I asked.

"Claro," said Alex. "My uncle died from a bushmaster." The venom is too powerful and hospitals too distant.

Then I had to say goodbye to my new friends. To be fair, I don't know if they'd consider me a friend so much as a very large outsider who asked too many questions about food (which may explain my size).

On the way downriver, we stopped at a village about an hour from Rurrenabaque. It was home to another tribe, called the Moseten. Like the Tsimane, they welcomed me with lunch—but of a different type. The primary ingredients stayed the same, but they breaded and fried the plantains and fish. I'd be lying if I said the Moseten lunch didn't taste better than Tsimane fare. But the researchers I spoke with discovered that the Moseten consume about 350 and 550 percent more sugar and oil than the Tsimane.

The Moseten are just as active as the Tsimane. Both tribes get about twenty thousand steps a day, which is something we should attempt to help our hearts, too. Yet research shows that Moseten hearts are beginning to look like mine—"they have poorer cardiometabolic health," said Gurven. As civilization creeps upriver, the researchers say that even some remote Tsimane bands are beginning to fry and salt their food and buy ultra-processed food and are, in turn, eating more.

This is something we see across the world, said Gurven. He pointed me to research on the Turkana people of northwest Kenya. Their traditional diet contains no ultra-processed foods. But more Turkana are moving to towns and cities. He explained that the Turkana who eat the traditional diet don't get chronic diseases. "But the Turkana who live in town—their diet has shifted to more ultra-processed foods and their physical activity has shifted as well," said Gurven. "And you start to see big changes in their diabetes and heart disease risk. And these changes can be rapid."

Back home in Las Vegas, I decided that for one month I'd eat as reasonably close to Leoncio's diet as I could.

Day one: I stepped into my pantry and took stock. There were 127 different foods. That is surprisingly fewer than the average American pantry. But just 15 of those were one-ingredient foods the Tsimane might eat, and I was counting canned vegetables.

The rest—even stuff we might traditionally think of as healthy, like whole grain cereal and Clif Bars—were packed with ingredients. My refrigerator was slightly better, with its fresh carrots and apples. But the walls of condiments lining the refrigerator door—caloric sugary, salty, fatty pastes designed to boost flavor—were a no-go.

I was left with rice, potatoes, beets, carrots, onions, apples. Luckily I had a freezer filled with lean elk from a recent hunt.

To supplement my diet, I drove to Costco, a place that, along with democracy, I consider one of America's great institutions. As I wove my oversized cart through wide aisles and throngs of people vying for free samples, I quickly discovered that most of this store was off-limits. Entire aisles and freezer cases held only ultra-processed foods.

My brain pinged off the bags and boxes of everything I craved but couldn't eat. My wife, meanwhile, dumped a train-car-sized box of Froot Loops into the cart. I viewed this as a hostile act, and the look on my face said as much. She looked at me. "Psshh. Sucker," she said, then proceeded to pace the aisles, seemingly making selections based entirely on the question, "What would Michael love to eat if he was not on this Tsimane diet?"

When I made my selections, she had all sorts of commentary. "Oh, so the Tsimane fish for . . . ," she said, grabbing an item I'd tossed into the cart and reading its label, "Trident Seafoods Wild Alaska Salmon Burgers?"

I wanted to keep this exercise as true to the Tsimane as possible. But I also had to be reasonable. Leoncio can walk down to the Quiquibey and cast a line for fresh fish. I live in a desert and must go to Costco for—yes—flash-frozen, deboned, and vacuum-sealed fish burgers.

Breakfast each day was oatmeal with berries and eggs. Lunch was rice or plantains, cabbage, and a salmon burger. Dinner was sweet potatoes, green beans, and elk. Dessert was an apple. I'd eat raw carrots or a banana if I needed a snack.

Despite its simplicity, my diet was far more varied than people throughout history would ever have consumed. Laudan, the food historian, explained that through most of documented history only a few foods dominated people's diets. "Every day you had the staple and perhaps a bit of meat," she said. Our abundance of choices is nice. But remember that people who have more choices of things to eat in a meal are more likely to overeat. It also explains why we often "have room" for dessert, even after finishing a big meal.

All our food choices have also made us persnickety at the table. Laudan said, "One of the problems of abundance is that everyone can design their own diet. In the past, if you didn't eat what everyone else in the family ate, then you just didn't eat. Nobody was going to make something special for you. This also led to a kind of family control on portion size in the diet, because there used to be enough food but not enough for everybody to have as much as they wanted," she said. "But today many children grow up believing they can pick whatever they want within reason. People aren't tasked to accommodate their tastes to anyone else anymore. There's an incredible individualization of our diets. This has ruined the sociality of eating to a very large extent."

With that in mind, I didn't mention my experiment when a neighbor invited us over for dinner. I ate what they served but preferred the foods that were most Tsimane-like, like salad and a bit of grilled chicken. I did my best at restaurants. Most places serve grilled meat with a carbohydrate like potatoes or rice. Sure, the restaurant portions were monolithic. And sure, they were all probably injected with enough butter and salt to kill a horse. But them's the breaks. This more liberal approach to portions and flavor enhancers like butter is why traditional restaurant foods contribute more to weight gain than fast food.

Along the way, my wife continued to treat my experiment like her pet comedy prop. When I'd cook dinner, she'd ask, "Oh, did the tribe use an Instant Pot pressure cooker to cook their meat?" Or, "Don't they drink water from the river? Maybe you should hike out in the desert and see if you can find water."

I couldn't help but laugh. And taking her guff was worth it. I began dumping weight fast. I went from 180 pounds to 175 in a couple weeks. The meals were satisfying. I'd found the zone of *enough*. I never fell into a food scarcity loop of mindless overconsumption. The Tsimane diet leveraged the second and third way we can get out of a scarcity loop. It took away unpredictable rewards and quick repetition. I saw it in myself, and Hall and Roussell saw it in their study participants.

I didn't, however, want to drop too much weight. To maintain 175 pounds, I had to eat more. And it was so much food. The Tsimane eat about 2,750 calories a day, but I had to eat about 500 more than that to maintain my weight. On my standard American diet, I'd always had the opposite problem, trying to not eat too much. But with the Tsimane diet, some meals felt almost chore-like. One day it took me thirty minutes just to eat the plantains I'd prepared for lunch.

But I was also sleeping better, my skin was clearer, my blood pressure and resting heart rate lower, I had more focus and was more productive at work, and I just felt better.

After a month, I'd finished my Tsimane diet experiment. But I'd become aware of a few things. Like how my diet soda habit was leading to heartburn at night. I stopped drinking it after 5:00 p.m. and slept better. Like the fact that air-fried plantains are a wonderful food—filling and delicious. Like how an apple for dessert can be nearly as satisfying as a bowl of Reese's Puffs cereal, for about four times fewer calories. Mainly, I found foods that didn't trigger the loop.

I continued most elements of the diet. My default meals became Tsimane meals. But I also didn't worry if I occasionally strayed and ate a cheeseburger and fries with a water-tower-sized diet soda when the occasion arose. I'm sure the Tsimane would do the same. Modern food is fantastic, we just need to balance it.

Most importantly, the approach was sustainable. An international team of researchers recently analyzed fourteen different diets and found that sustainability is critical. As the brilliant nutrition scientist Layne Norton, PhD, explained, "What this analysis showed is that we need to ask, 'What is going to be the easiest diet for you to adhere to in the long term?' And you should probably do that."

At the end of our conversation, Gurven and I were talking about how we generally know what's good for us: single-ingredient foods, exercise, and having less stress rather than more. But the diet industry has made health more complicated than launching a satellite into orbit. It's pushed us into strange diets and granular prescriptions around exercise and sleep and more.

"I think we shouldn't get too stressed over any one little thing," he said. "If the details of your healthy eating and exercise and sleep routine create anxiety, you're probably doing it wrong. And let's say some complicated health prescription does add an extra year to your life. Well, you have to subtract all the painful time and effort you spent worrying and doing that thing that stressed you out."

Five pounds lighter, I was ready to explore what else scarcity brain craves. I threw too much stuff into the back of my car and headed to grizzly country.

Stuff

After my thirteen-hour drive from Las Vegas to Missoula, I was finally a passenger. Laura Zerra was driving us east in her 2017 Jeep Cherokee across Montana's Highway 200. It's the longest state highway in the country, a great strip of two-lane blacktop that runs horizontal across Big Sky country.

Towering pines flanked the asphalt. They were intermittently shading the road, creating a strobe effect as we moved through sun and shade. It lit and dimmed the thick black ink covering Zerra's arms and legs. Her tattoos are like some fever dream of Carl Jung and Joseph Campbell. All myth and symbol.

On her right forearm there's an ouroboros snake, the symbol of death and rebirth seen across ancient cultures. A Navajo spider on her left forearm. A Zuni crow on the inside of her upper right arm. There's a werewolf eating an old lady, which she pulled from a woodcut from the nineteenth century, on her right outer arm. On the opposite upper arm there's a hyena.

"Everyone thinks hyenas only scavenge," she told me as she piloted the Jeep around a bend. "But hyenas actually hunt more than lions. I love misunderstood animals." Which explains the vulture on her inner arm.

"Scavengers like vultures get a horrible reputation. But they play an important role in the ecosystem. And I can identify with that," she said. "I relate in a very real way to living off the scraps of everyone else's excess."

Modern humans are also animals. Sometimes our smartphones, Doritos, and carpeted homes make us forget that. But some of us live closer to our original form. Those people tend to be like hyenas and vultures: misunderstood. Like Zerra.

We'd left Missoula an hour earlier and had reached the outskirts of Lincoln, the town where Ted Kaczynski, the Unabomber, lived in his off-grid, ten-by-twelve-foot cabin. Zerra was leaning back behind the wheel and telling me, "I was sort of an oddball kid. I loved spending time in the woods. I loved the animals out there and wanted to see them up close and personal. I ended up befriending and spending most of my time with a pack of coyotes, or really just teaching them to tolerate me. And those coyotes . . ." Her voice trailed off.

She pulled her foot from the gas pedal, slowing the Jeep. Her back straightened and she took the form of a vulture, eyeing the shoulder of the road.

A mule deer lay on the road's shoulder. Roadkill. Zerra craned her neck and squinted as the Jeep continued its deceleration. Slower. Slower. Slower.

"Ahhh, nope," she said, plunging the gas pedal to reaccelerate. "This time of year, in late spring, if you don't find roadkill super early in the morning, it's going to be no good."

Zerra carries what she calls her "serial killer kit" in the back of the Jeep. It has a bone saw, a tarp, gloves, scalpels, a rope, and trash bags. Had the deer been killed recently, we would have pulled over, butchered it, then driven all the meat back to Missoula. "I hate the idea of animals dying for a bad reason," she said. "The meat doesn't have to go to waste."

Zerra's dog, a Belgian Malinois named Nerron, eats only fresh roadkill. "I'm at the point now where, even while driving, I can easily tell how long ago the animal was hit," she said. "I used to have to pull over to check."

Most of what Zerra owns is in this Jeep. There's the serial killer kit, of course. But also a big Eberlestock roll-top backpack half-filled with very specific gear.

Before we set off, we were at the home of her friend Jana Waller, a celebrity in the hunting world who hosted a show on the Sportsman Channel and now CarbonTV. Zerra in 2021 started living in a guest bedroom of Waller's home, because she liked the place's easy access to Montana wilderness.

Zerra and I were sitting on the living room floor with all of our gear laid out and were preparing to spend a few nights in the Bob Marshall Wilderness. It's a million-acre strip of roadless Montana wild that the U.S. Forest Service says contains one of the highest densities of grizzly bears in the lower forty-eight states. But our gear piles for the impending trip looked different.

Hers contained the following: a tarp that she uses for a shelter, because it's far smaller and lighter than a tent. An "oh shit bag" filled with an emergency blanket, compass, lighter, fire paste, and Super Glue for gluing together gashes. Knife. Rain pants and jacket. Toilet paper and toothbrush. A warm fleece and down jacket. A headlamp and tin saucepan for cooking. Bear spray. A freeze-dried backpacking meal for each night. A loaf of bread and three sticks of butter for breakfast and lunch. A spoon.

"Normally I wouldn't bring this," she said, holding up the spoon. "I'd just bite the stick of butter and spit it on the bread. But I didn't want to do that in front of you. The spoon is so I don't horrify you."

I looked at my own pile. My clothes appeared as if I planned to be some outdoorsy Diana Ross. Multiple socks and underwear and layers. I also had a roomy tent. A first aid kit that could treat all the wounded on Utah Beach. An assortment of backpacking dinners, breakfasts, candy bars, protein bars, and the like (the Tsimane experiment was clearly over). A knife. A multitool that also had a knife and thirty-seven other functions. On and on. And I planned to carry it all in a backpack the size of a sedan.

"Do you think I have too much stuff?" I asked.

"Ummm . . . you need to find what works for you," Zerra said. This seemed like a nicer way of saying, "You have wayyyyyy too much shit."

Looking to learn something from Zerra, I began paring down my gear pile to look more like hers. I swapped my big backpack for a smaller one. I ditched my tent and borrowed one of Zerra's tarps. "In bear country, you're just trapped in a tent," said Zerra. "Tents are like a fabric prison."

I left behind extra layers and socks and underwear. I resisted the pull of my fat kid genes and culled my food.

Zerra had placed an extra canister of bear spray and a Glock 10 mm handgun on a nearby table. Regarding grizzly bears, she said, I could bring one of these items, both of them, or none of them. My first impulse was to heavily arm myself. I'd need the Glock . . . yes, of course . . . and perhaps Zerra also had a Hellfire missile lying around.

Then higher-order thinking kicked in. Carrying the Glock, I realized, assumes that I would be able to draw, fire, and place a 10 mm bullet in the eight-inch-wide forehead of a grizzly that was bearing down on me like a Formula 1 car. And this would all need to happen as I was shrieking and emptying my bladder and bowels while also regretting that I stopped going to church. This seemed like quite the technical ask.

So I, too, brought only the bear spray. I was confident I could at least send up a cloud of it at a charging grizzly. Enough spicy fog to blind my vision so that I wouldn't have to watch my thrashing.

"Good," Zerra said as I attached the spray to my pack's hip belt. "With that gun, you'd probably miss or just hit it in the body and piss it off more," she said. "Or if the bear was mauling *me,* you'd probably end up shooting *me* in the head." It's nice to have people who believe in you.

My pack was still stuffed despite the culling. And I had one final item I wanted to bring: an MSR WindBurner Backpacking Stove. I worked all sorts of different angles trying to fit it.

After a few minutes of jostling and wrenching on zippers, I gave up. "We can just use your pot to boil water," I said. "My stove was just for coffee anyway." She must have caught a trace of sorrow in my eyes, because Zerra grabbed my stove. "I'll carry it for you."

Serial killer kit aside, Zerra is shockingly sane.

"I don't know how to put this," I told her back on the road, as the Jeep accelerated past the deer's corpse. "But if you look past the roadkill thing, you're pretty normal for a person who does what you do."

She laughed. "And this is why I'm not really a survivalist. Or, at least I don't consider myself one," she said. "I can just live with nothing, and have for a long time."

Spoon notwithstanding, Zerra has lived out of some version of this pack for the last fifteen years. And she's being modest. She's definitely a survivalist. But not the theatrical kind, of which there are various flavors. There are the mountain man survivalists, who dress up in beaver pelts and carry around muskets at frontier revival gatherings. There are the prepper survivalists, who ready themselves with skills and food caches for an impending nuclear apocalypse or alien invasion. There are the extremist survivalists, who are mostly concerned with living off the grid and hoarding guns and bullets for the rise of a tyrannical government. Then there are the primitive skills survivalist, hippie dream-catcher types who grow and make their own food, clothes, and homes and just generally try to be "one" with Mother Earth.

Zerra is a one-off creature. A runner, gunner, explorer, traveler, and, generally, seeker of the human experience in all its varieties. She's traveled the world with nothing to her name for years. But she also did a short stint traveling with billionaires. She's made money as an offshore fisherman, butcher, farrier, mushroom forager, and farmer and also appeared on the show *Naked and Afraid*, where she won so easily the producers kept asking her to come back for even longer, tougher challenges. She once even won a car on *The Price Is Right*. The human experience in all its varieties, indeed.

Zerra wasn't always so unconventional. "I was always good at school. I could write papers and regurgitate information. I planned on being a doctor or veterinarian," she told me as she kept pushing the Jeep. "I was told the way you got happy and lived the best life was to get good grades so you could get the best job. But what I really wanted to do is explore different environments and learn how to survive in them." She figured becoming a doctor would be a good way to hoard cash so she could explore and test herself in the wild during her paid annual time off.

She entered Connecticut College in 2003. It's one of the more elite liberal arts colleges in the country. Part of the "Little Ivies" on the East Coast. She got straight As. But even then she filled her free time doing what she loved. Her dorm became a place to shower and store books, and she mostly lived out of a shelter she built in the Connecticut woods near the school.

But sometimes it takes the worst of the human experience to lead us into the best of it. "I had a really good friend who was amazing and smart and talented," Zerra told me. "But she was so stressed out by exams and work. This was because she also thought that the way you live the best life is to get good grades to get a job so you could do what you want on your weekends and vacations. My friend got so overwhelmed that she tried to commit suicide in college. I actually walked in and found her. It switched something in my brain where I wondered, what's really important? Everything in my life had built up to getting this piece of paper. I realized that I wasn't learning anything in class I cared about. The information wasn't applicable for what I wanted to do with my life."

A person can, she realized, travel and explore different environments and learn how to survive in them for free. For as long as they want. And they don't have to worry about vacation days if they don't have a job.

So she quit school with a semester left. "This was the thing I did that people were most upset about, but that I was most proud of," Zerra said. "Because it was the first thing I ever did for myself and only myself and not based on the narrative that everyone else was holding themselves and me to about the way the world works. I had my own back. I believed in myself."

She packed an older version of the same gear bag and started traveling. And there I was, with her at this moment in her journey.

I thought Zerra might have something wise to teach me about living well in our world of mass consumerism. If the data is correct, impulse buying spiked during the pandemic and has remained elevated ever since. I was still processing my own behavior.

Online algorithms seemed to put items perfectly tailored for my personality and interests in front of me when I was most prone to distraction. And I'd buy.

For example, during the pandemic quarantines, I had nowhere to go and nowhere to be and a clock on my phone, which I looked at every five minutes. Yet I bought a ridiculously expensive luxury watch. I'd wear it around the house while dressed in $14 sweatpants from Target and a T-shirt I got for free from my dentist.

Beyond mindless online shopping, I had one particularly dishonorable episode. I've tried to repress the details of what happened, but it ended with me trading in my beloved, four-year-old pickup truck for a Subaru Outback.

It wasn't just a pointless purchase given that my truck still had plenty of life left. But from a personality standpoint, it was a terrible fit. Nothing against sensible family vehicles like Subaru Outbacks, but I don't have kids and a granola hatchback isn't exactly my vibe. It was the equal opposite of a soccer parent trading in their Honda Odyssey for a Harley-Davidson Shovelhead.

"You bought a Subaru . . . Outback?" my wife asked when I got home. She was looking at me the way you might an otherwise loyal and well-trained dog who has just urinated all over your favorite rug. It was a look of disgust and disappointment—but also genuine concern that this animal could be deeply unwell.

"The car is a Consumer Reports Best Buy!" I exclaimed.

She just shook her head. It wasn't just that my impulse control seemed to be in pieces. It was also that my newly unhinged impulses were directing me into baffling new territory. It was like the prodigal son coming home buzzed—on angel dust.

It felt like the scarcity loop. I'd think of or see a product that I thought might improve my life. Then I'd search the internet for the right version of it and eventually stumble upon a winner. Then I'd repeat the cycle. The UPS lady and I were on a first-name basis.

But because my new possessions accumulated and accumulated, they seemed to begin possessing me. For example, I was taking consulting gigs I didn't necessarily want to, but felt I should to cover the ensuing bills and rationalize a purchase.

Then, once I realized I had too much and that started to annoy me, I'd hit a tipping point and go searching for answers. I couldn't ditch the bills. But what I found online suggested that I'd be better off if I just purged my stuff. I was guided into less, or "minimalism."

But I never felt as if any of these resources helped me ask the more fundamental question about why I was buying too much in the first place. Tidying up felt like temporary relief. It didn't break the cycle; it was *part* of the cycle.

So I wondered if Zerra might have a solution for this loop of buying stuff. I had no delusions that I'd soon be living out of a backpack and selling my air-conditioned storage unit I call a home. Nor did I want to.

Rather, I wanted to think deeply about what my own version of Zerra's backpack might be and how that might change me. Maybe by looking at the deeper why behind my possessions, I could find enough.

It's nice to think that our ancestors were unattached, enlightened beings who lived in societies akin to hippie communes. As if possessions weren't important to them and everything were all kumbaya and sharing and peace and love.

While it's true that our ancestors didn't have as much as we do today, the idea that there have ever been people today, yesterday, or even millions of years ago who didn't care about stuff is a fairy tale. "All people are materialistic to some degree," wrote scientists in the academic textbook *Consumer Behavior.* "People naturally yearn for more of whatever material resources are prized within their culture."

Indeed, despite feel-good, Disney-movie ideas about our ancestors being "one" with nature, taking only what they needed, the truth is often quite the opposite. Nature was often brutal and much more like a Tarantino film. To survive, we often had to be equally brutal. And it's nearly always been better for our survival to have too much stuff rather than too little.

Take, for example, the practice of piskun among the Blackfoot tribe. It translates to "deep blood kettle" and describes a two-thousand-year-old hunting practice where tribe members would chase herds of buffalo. They'd eventually form them into a funnel and run the herd off a cliff, killing hundreds of animals at once. It was the Blackfoot equivalent of Thanksgiving and Black Friday: a massive event of acquiring food and stuff.

They rendered thousands of pounds of meat and material goods from this. The buffalo skins built shelters and clothes. Bones and other parts made good tools and military equipment. But the tribe members couldn't render all the buffalo quickly enough. A significant amount of the meat and materials rotted.

That's just one example. Ancient human societies all differed in how materialistic and sharing they were. None, however, were totally egalitarian. Many were pure hoarders. Like the Kwakiutl of Vancouver Island. The anthropologist Helen Codere, who studied the tribe, wrote, "Each household made and possessed many mats, boxes, cedar-bark and fur blankets, wooden dishes, horn spoons, and canoes." She explained that they just kept making stuff. "The production of more of the same items was [never] felt to be superfluous," she wrote. Like, why have just one horn spoon for each person when we could have fifty horn spoons per person? This practice began long before Europeans arrived, so it's not as if they got the idea to hoard from elsewhere.

Or there were the tribes south of the Kwakiutl. Thanks to the climate and bounty of what is now California, tribes in the area rarely had to move. And so they spent their time piling up stuff. The scientists wrote that they had an "obsession" with wealth accumulation.

Anthropologists at the University of Texas believe our default is to collect more stuff rather than less. They theorize that there are three reasons humans evolved to love material goods.

The first is that having stuff helps us survive. The scientists explained, "In the past, owning the right goods at the right quantities provided protection, comfort and greater capacity to trade for other needed goods." It's probably best to think of these items as "gear" rather than stuff. Gear has a clear utility. Gear like tools, shelter, and weapons helped us accomplish life-giving tasks. The same rule applies today.

Zerra's relationship with her possessions is the ultimate expression of seeing items as gear. "Everything I own must earn its weight," she told me. "It has to serve multiple functions."

The second reason we like material goods is that they can bring us status. Having the right stuff for our time and place can boost our social ranking. Certain goods serve not only their primary purpose but also a secondary purpose of telling others about our rank or place in society.

The American economist and sociologist Thorstein Veblen in 1899 called this "conspicuous consumption." It's when we buy expensive stuff over cheaper yet functionally equal stuff to flex our social status. He wrote, "Conspicuous consumption of valuable goods is a means of reputability to the gentleman of leisure." For example, no one *really* buys a Cartier watch to learn what time it is. Nor do they buy a Louis Vuitton bag just to carry their crap.

Scarcity and exclusivity are the secret sauce of luxury brands and selling stuff at a premium price. Scientists at Temple University discovered that scarcity cues, the belief that an item is hard to get or limited, outperform popularity cues, like advertising that an item is a best seller.

Conspicuous consumption and hard-to-get goods enchant even the most enlightened among us. Case in point: His Holiness the Dalai Lama owns fifteen luxury watches. The collection includes two Rolexes and one exceedingly rare Patek Philippe. For those of us not in the horological know, the latter brand is what you buy when a Rolex feels too cheap.

Third, we can use material goods to feel as if we belong. This is different from buying stuff to get status. The scientists explained, "The motive for status manifests itself in efforts to be *above* others within a group, whereas the motive to belong manifests itself in efforts to be *with* others in a group."

Scientists now call this "brand tribalism." It's where we find social meaning from the purchases we make. Consider brands like Whole Foods, Goop, Black Rifle Coffee Company, and Patagonia. Shopping and displaying the logo is as much a sociopolitical, in-group, near-religious act as it is buying food, supplements, coffee, or an overpriced puffy jacket. Humans have been stamping brands on items for at least five thousand years.

But for nearly all of time, we could have only so much stuff. The materials we needed to make our stuff were harder to extract, and shaping our goods by hand took much longer. For example, it used to take a blacksmith one minute to forge a single nail. Nails were so scarce and valuable that arsonists would often burn down buildings just to steal nails. Today, modern nail-making machinery can crank out 360 nails in one minute.

Another reason we had less in the past is that many groups of people were constantly on the move. They often abandoned items that were hard to carry.

So it's not that we wanted less in the past. We craved more and gladly took it when the opportunity arose. But the opportunity rarely arose.

Even in the eighteenth century, most Americans didn't own much. A shelter. Some tools that helped them acquire and cook food. Some basic furniture. A Bible. A few items of clothing. Some nails depending on how many buildings we'd recently burned down. American men and women back then, for example, had an average of three outfits. Even the richest people didn't need walk-in closets. Thomas Jefferson's wife, Martha, for example, owned seventeen outfits.

The third part of the scarcity loop is quick repeatability. In the past, purchasing couldn't fall into a scarcity loop. We were too poor and stuff was too scarce and expensive. People were far more likely to make or buy something once and use it until it was completely worn and unusable.

But the eighteenth century marked the beginning of the end of that. In 1733, the Englishman John Kay invented a weaving machine. It doubled the rate at which a textile worker could weave.

But that didn't just mean we started getting more textiles. Kay's invention began a domino effect. It increased the demand for yarn. To keep up with the demand, we realized we needed to make a machine that could make more yarn. This then meant we needed a machine to get more cotton. Then a machine to make more machines. And on and on. One machine would increase the productivity of one industry so much that it would spur other industries to invent or adopt new machines to supply demand. And so on down the line until all industries were cranking out materials thanks to mechanization.

By 1850 a full industrial revolution was taking hold. As we began figuring out how to make stuff faster and cheaper, our environments of scarcity, which we'd been in since our genus *Homo* split off from chimpanzees about 2.5 million years ago, began shifting to those of abundance.

If we were to open our closets to Martha Jefferson today, she'd be shocked. The average American purchases 37 items of clothing each year. One study found we now own 107 items of clothing. That study also detailed how we feel about those 107 items. It discovered that we consider 21 percent of those clothes unwearable. We think 57 percent of the items aren't great—either too tight or too loose. Then we have an average of 12 percent we've never worn. And that leaves us with a Martha Jefferson–esque 10 percent—11 items—that we regularly wear. The EPA says we throw away about sixty-eight pounds of clothing and textiles per person per year.

But mechanization, of course, didn't just give us more clothes. It gave us more of everything.

Today the International Shipping and Packing Association, the trade group for moving companies, says the average American home contains about ten thousand pounds of stuff. That's spread across anywhere from ten thousand to fifty thousand items light and heavy—from pens to TVs. The *Wall Street Journal* found that Americans now spend $1.2 trillion annually on stuff we don't need.

Psychiatrists first started noticing compulsive buying among the rich in the 1930s. But now we don't have to be rich to buy compulsively. The historian Jeannette Cooperman wrote, "Only in the twentieth century did people begin engaging in the eccentric over-accumulation of random, not terribly valuable stuff."

A study in *Frontiers in Psychiatry* found that 6 percent of Americans suffer from compulsive buying disorder. Some studies suggest the number is 2 percent, while others believe it's as high as 16 percent.

The prominent American psychiatrist and University of Iowa professor Donald W. Black identified distinct phases compulsive shoppers go through. And they echo the scarcity loop. First the shoppers look for good opportunities—places with frequent sales or new items. Next is searching: they enter into the world of stores and malls or online shopping sites. Black noted that these shoppers usually look for unpredictable bargains. It could be a generic handbag selling for $30 instead of $60. Or it could be a luxury bag discounted for a limited time by 40 percent, down to $2,200. And so they buy it—then quickly start searching for the next item to buy.

Black wrote that people with compulsive buying disorder "describe [shopping] as intensely exciting." And like addiction, it's often a learned coping mechanism, wrote Black. "Negative emotions"—for example, depression, anxiety, boredom—"were the most commonly cited antecedents to [shopping], while euphoria or relief from the negative emotions were the most common consequence."

Black said our abundance of cheap items means the disorder affects low-income people just as much as it does rich people. And the quick repeatability is key, he explained: "Many compulsive shoppers buy in quantity."

Even our concept of "need" has changed. We like to think that necessity is the mother of invention. But the Pew Research Center pointed out that it's actually the other way around. Invention is the mother of necessity. We invent something we don't exactly need, but over time it begins to dominate our culture so much that we start to consider it a necessity.

Like GPS, bluetooth, wireless headphones, remote controls, television, microwaves, and on and on. We shape our machines, and our machines shape us in a way that makes us increasingly reliant on them. The number of items Americans consider "necessities" rather than "luxuries" multiplies each decade.

As I read these figures, they seemed too high. Absurd, even. Who owns ten thousand items, much less fifty thousand? And then, while writing this paragraph, I paused and counted nineteen items on my desk alone. Pens, the container for the pens, a coffee mug, room spray, computer, cell phone, monitor, webcam, adapters, desk lamp, books, a Kindle, a notebook, and more. As for my entire office? LOL. The unnecessary highlights: a Camel cigarettes sign in Arabic from the Middle East, far too many books (most half read), a Jerry Garcia mug shot on the wall, and a bottle of some strange powder I purchased off a guy in an alley in Bangkok. He told me to use it "only in emergency." In truth, if an emergency struck, there was only thing in my office I'd take with me: my dog, who was curled up on a Tempur-Pedic dog bed, surrounded by three dog toys.

Yes, even many dogs in the United States now have more possessions than Americans just a few hundred years ago. My wife and I don't dress up our dogs, because animal cruelty is wrong. But the dog clothing industry alone is projected to be worth $16.6 billion in the next handful of years.

Unlike our ancestors, we aren't forced to occasionally leave our stuff behind. To radically purge it. We cycle through and collect things across our lives. As Mary Oliver put it in her poem "I Own a House," "I own a house, small but comfortable. In it is a bed, a desk . . . a telephone. And so forth—you know how it is: things collect." Things collect, indeed.

But our things aren't collecting in small cottages, like the type Oliver lived in. The average home has grown 75 percent since 1910. It is now roughly twenty-five hundred square feet. In some cities, homes have tripled in size, like in my hometown of Las Vegas.

As industrialization swept the globe, our scarcity brain kept doing its thing, pushing us into more. We now all have so much shit that we have an entire industrial complex designed to help us manage all our shit. We buy books and watch TV shows that teach us the magic art of culling and organizing our shit. Storage units, discrete locations we pay for so we can hoard more and more shit, are not only a thing; they're also one of the nation's fastest-growing business segments. There are now more self-storage facilities in the United States than McDonald's, Burger Kings, Starbucks, and Walmarts—combined.

UCLA scientists say part of the reason we tend to collect so much stuff is that we don't have a biological governor that tells us we've over-bought. If we're eating or drinking, we frequently overdo it, but eventually we'll fill up and have to pause. But this isn't so with stuff (or influence or information, for that matter). We can always repeat. Quickly. And get a storage unit.

Here's a good rule of thumb to help you decide whether to buy something new or donate an old item: decide within sixty seconds. The psychology researcher Melissa Norberg, who is the president of the Australian Association for Cognitive and Behaviour Therapy, wrote, "Whenever you find yourself taking longer than a minute to make a decision, it's likely you are trying to find a justification for making an unnecessary purchase or keeping an unneeded item."

Many of us feel guilty about our purchases. We believe our immensity of stuff is killing the environment. Like, each Amazon Prime purchase or $6 T-shirt we buy from some fast-fashion house is a vote to kill the polar bears.

But I spoke with Andrew McAfee, who lifted some guilt—albeit with two caveats. McAfee is the co-director of MIT's Tech for Good

Research Group. He explained that until the 1970s it appeared the earth was headed to environmental ruin.

He told me, "Pollution was going up. We had wiped out species, and ecosystems were under threat. Resource use was going up exponentially year after year in lockstep with overall economic growth." But sometime around the first Earth Day, resource consumption began to decline, even as our population grew. This trend continues and accelerates every year. We get more people. Our economy gets stronger. But we use fewer resources.

"We finally figured out how to lighten up on our planet, even as we continue to grow our populations and economies," McAfee said. And it's not that we started consuming or desiring less. The opposite is true.

"Rather," McAfee told me, "there's been a one-two punch of competitive capitalism and a very good and evolving technological tool kit that began to allow us to do more with less. And this is for the very simple reason that molecules and resources cost money and companies like to save money. So when opportunities via technology come along for them to do more with less, to get their goods or services out there while spending less money on materials, they'll take that deal all day long."

For example, in 1960, a single can of beer used more aluminum than an entire six-pack of beer does today. Aluminum cans went from weighing eighty-five grams to just under thirteen. Beer brewers probably didn't consider the world aluminum supply. They did, however, consider their bottom line. Using less aluminum per can saved beer brewers (and drinkers) money. This general rule holds for most resource use.

In the United States, levels of six air pollutants including carbon monoxide and lead have declined by 77 percent since 1970. Meanwhile, our GDP and population grew by 285 percent and 60 percent, respectively. We've reduced our total carbon dioxide by 13 percent since 2007. That's thanks to simple economics and smart government environmental regulations.

Not only are we using resources more efficiently, but we're also seeing old innovations merge into one. For example, in one of my lectures, I show students a famous RadioShack advertisement from 1991. It featured all sorts of products on sale at the store: a computer, a telephone, a clock radio, a stereo, a calculator, a police scanner, a camcorder, a camera, a voice recorder, an answering machine, and more. And the original advertisement was, of course, printed in a newspaper.

"All of those items have now vanished into smartphones," explained McAfee.

Put these two phenomena at scale and combine them with wise government interventions that reduced pollution and protected wildlife, and we've been able to improve the human condition while treading more lightly on the planet. We're now consuming fewer resources, using less land, polluting less, and even bringing back species we'd nearly pushed to extinction. All as our population grows and living standards rise.

McAfee said he'd bet that in a decade we'll be using fewer resources than we are today, regardless of how much our population and economy grow.

That said, we're not entirely off the hook. For one, McAfee told me, "global warming is real. It's caused by us and it's bad. There are also parts of the world where there is too much pollution and it's getting worse." He believes the answer is to do what's worked since 1970. "We know the playbook for dealing with those issues," he said.

World population, he pointed out, is projected to begin to decline around 2050. "Our planet is abundant enough to satisfy our consumption," he told me. "It's counterintuitive because it's been drilled into us that we're ruining the planet and it's getting worse and worse." But the data suggests we'll be okay if we continue innovating and spreading those innovations worldwide.

The second problem is one that advancements like a better aluminum can and having a bunch of items merge into one can't solve. It deals with the human condition. As we've become more comfortable and adopted more efficient technologies and stuff, we haven't necessarily become more satisfied. Mental health issues are rising around the world. Many of these technological shifts are, in fact, causing our malaise. They're disconnecting us not only from others but also from ourselves and ways of living that satisfy us.

"Abundance brings with it problems," McAfee told me. "We should prefer those problems to the problems of scarcity, but they're still problems."

Zerra spends six months a year in the wild, far off the trails. But she's not a casual camper or nature ambler. We'd been cranking into piney meadows, over rocky knolls, up boulder fields, and through thick brush and trees.

It was the afternoon of our second full day in the backcountry. Perhaps twenty minutes earlier we'd paused to eat "lunch." For me it was a mint chocolate chip Clif Builders bar. It tastes as if you took a sip of chocolate milk after brushing your teeth.

Zerra ate butter. She used her spoon to scoop a quarter of a stick of butter onto a slice of bread. She then folded it like a taco. A butter taco.

The problem is that after a butter-forward lunch, Zerra is like a young terrier who has gotten into espresso beans and now has the zoomies, running laps around the home at full tilt. Except Zerra and I were in some of the most rugged mountains in the lower forty-eight states. Going higher and deeper into wilderness.

There is hiking. And there is what Zerra was forcing me to do. Covering ground that even some of the most sure-footed four-legged animals would avoid. She was out ahead as we scrambled our way up a rocky slope.

She pointed to the top of the slope, which ended in vertical cliffs that shot hundreds of feet up. "It's usually flat right at the bottom of the cliffs," Zerra said. "Rams hang out in places like that a lot. They often die there. So that's prime territory to find a skull."

In the fall, Zerra sometimes hunts for meat. In the spring and summer, she's all about shed hunting—essentially searching for and collecting antlers left naturally by the animal. Each winter, ungulate animals like deer and elk shed their antlers. She sometimes even finds the skulls and bones of deer, elk, and bighorn rams.

Zerra prefers shed hunting because, she said, "I can do it any time of year. It's harder because the animal isn't moving and I'm looking for something smaller. I have to be far more aware . . . I'm not really looking for antlers. I'm looking for a moment in time, a moment in time when an animal died, usually so another could live."

Some people shed hunt as a lucrative side hustle. Collectors will pay thousands for a perfectly preserved ram skull with thick swooping horns, or the antlered skull of an elk. Antler is one of the strongest natural substances on earth for its weight, and well-preserved antlers can be used for knife handles or as decoration on any other item an outdoorsy Etsy seller can dream up. But even weathered antlers can be sold to companies hawking them as dog chews for around $30 a pound.

"I don't sell the skulls, horns, or antlers ever," Zerra told me. "That would totally ruin the search for me." The pursuit is entirely for the thrill of the highs and lows and clear goal of searching, which captivates her and motivates her to stay in the wilderness for longer. I pocketed this information. It sounded a lot like the scarcity loop ideas of Daniel Sahl and Thomas Zentall.

We eventually reached the top of the slope. We craned our necks to look up the cliffs above us that only went higher. "This looks like the perfect place to die," she said. It was craggy and protected up there. A place only a handful of animals would ever want to go.

Then she hustled along a flat space just below the cliffs. She was like a bird dog who'd picked up the scent of a pheasant, scanning up and down the pathway as she rapidly covered ground. Eventually, the flat cliff bottom shifted to a steep slope that fell thousands of vertical feet.

"No luck," said Zerra. The place rendered nothing. It was a near miss. But we did get a hell of a view. The sky was bright white, punctuated with dark gray clouds that looked as if they might open up on us. The lake below wound like a fat serpent, slithering between mountains. We sat for a moment.

Zerra told me that during her travels she was forced to become like some hobo wilderness MacGyver. "I had some pack I got for cheap at a military surplus store. In it was a wool blanket, a water bottle, a pot, a knife, a journal, my passport, and a Bic lighter." She'd have to get creative if she didn't have something. "I found this piece of Tyvek and it worked great as a tarp. I would almost get a high from being in a situation where I'd have a problem and would have to get creative to figure out the solution. Because I couldn't just go buy something. This opened me up to being completely present in my experiences, and it was just so rewarding."

Zerra might sound kooky, but she's alluding to something profound.

Scientists at the University of Illinois and Johns Hopkins recently wrote, "Consumerism and over-acquisition have become the order of living and abundance has emerged as the norm, especially in the [developed] world." The scientists say that because we have ample access to all kinds of resources, we default to solving problems by buying.

To understand the downsides of our buy-to-solve tendency, the scientists took two groups and conducted six different experiments. The first group was told resources were scarce, while the second group believed resources were abundant.

Each of the six experiments asked the groups to creatively solve problems with the resources they had. For example, in one of the experiments the participants had to come up with as many uses for a brick as they could. In another, they got a handful of Legos and were asked to devise the coolest toy possible. In yet another, they received a candle, a pack of matches, and a box of tacks. They had to figure out how to jury-rig the candle to the wall without having it drip wax to the floor when lit.

In all six studies—all six!—the participants who faced scarce resources performed better. They came up with more uses for the bricks. Their toys were the most fun. They morphed into MacGyver. Overall, not only did the scarcity groups come up with more solutions, but their solutions were also more efficient and creative.

The takeaway: when stuff is abundant, we tend to fix any problem with more stuff. To buy and add. We're more likely to use items as advertised because there must be some other gadget out there we can buy to solve our problems.

The opposite happens in scarcity. Yes, our first tendency is to solve our issue by piling on more. But humans are persistent, creative creatures. We don't give up if we can't solve a problem with adding. Our species would have died off long ago if we quit when we didn't have enough. In the modern world, if we push back against our tendency to add—forcing ourselves to solve a problem with what we have—we'll likely solve it better, more creatively and efficiently. Creativity and efficiency bloom under scarcity.

This study reinforces years of research going back decades. By facing constraints, we often end up accomplishing more. And this isn't even a relative "more with less" thing. It's just a more thing.

Zerra found that her experiences and aha moments of solving problems with limited resources also changed her worldview.

"I was not a people person when I started," she told me. "I enjoyed being alone. I mostly thought that other people were just ruining the wilderness I loved so much." She assumed she'd bum around for a while, then settle into a hermetic lifestyle in some off-grid cabin.

"But I started to fall in love with people through hitchhiking," she explained. "I'd get in a car and have a finite amount of time with a person.

And we both knew we'd never see each other again. People would be completely real with me because they didn't have to pretend to be who they thought they had to be in their normal lives. People would tell me their deepest, darkest secrets and we'd be crying with each other. It was beautiful experience after beautiful experience. And that came from the freedom of being able to be completely in the moment and untethered. I didn't have to worry about working to get, say, $400 for a plane or bus ticket. That would have totally changed my experience and altered those hours of my life. I would have gotten on the plane or bus, not talked to anyone, and arrived at my destination. I would have missed out on so many experiences that completely changed my life and perspective on the world."

After we'd taken in the view for a couple minutes, Zerra was moving again. We began descending, surfing the rocks downhill. The sliding rocks were loud enough that she was yelling.

"Hearing people's stories and just letting them talk. I think that's when I got the idea that I really just wanted to experience the entirety of the human condition. The ins and outs, ups and downs, goods and bads of it and just have a really rich experience of living," she yelled through the thin Montana air. "I didn't think these human experiences would be all good. But I thought they'd all be important. They'd change my perspective. Because I could see that everyone is just doing the best they can with what they have, you know? Generally people are good. But we all fit the mold we cast. If we focus on the negative, we're going to look for negatives. But if we approach things with even a small amount of compassion and real interest and give people the benefit of the doubt, we can see how people don't fit our narratives. We find that people are amazing."

It's important to note that Zerra is not some trustafarian. Her parents are solid middle class—a preschool teacher and an electrical engineer. "I got my first job at a farm when I was fourteen. It was probably illegal work," she said. "But my parents told me that if I wanted something, I had to pay for it. And I wanted a car when I turned sixteen so I could head to the wilderness."

Even though Zerra dropped out of college, she was still on the hook for loans. Big ones. She worked odd jobs for a month or two during her travels to cover her loans for the year.

She also did a stint living entirely counter to her typical vagabond life. Zerra's bit of TV fame led her to befriend a group of mega rich who enjoyed playing at roughing it. They'd all go, for example, on $40,000 hunting trips.

Zerra said the psychic weight of money was always there. "I noticed that the more people had, the less involved in the moment they seemed to be. They were more involved in the future. Doing and maintaining stuff and everything that came with it.

"Everything would be prearranged and planned and scheduled," she said. "I describe it as a really expensive Happy Meal. It was a carbon copy of what every other rich person got. Things went exactly as planned. And it was just so different than the experiences I had when I had, like, six bucks and a backpack filled with a few items and I'd go out into the wilderness and have to figure things out myself. Those experiences were so unique, and I just didn't get them when we had all this money to have everything perfect."

As the slope let off and the rocks transitioned to pines, Zerra was scanning for antlers and skulls. She told me she understands that working people face constraints. So this isn't any sort of judgment. "If you were running a company or had a lot of family counting on you and wanted to get away but only had six days," she said, "I can understand why you'd pay to get the most out of those six days. You'd end up buying the Happy Meal experience. You don't want to plan and you don't want anything to go wrong. You have limited time and you're thinking of all the other responsibilities you have.

"I get that," she said. "But I just feel like money hasn't served me in the sense of having better experiences compared to not having as much money. Money brings more control but also less adventure. When I start with fewer resources and materials, whatever I earn, whatever problem I solve myself, I enjoy every moment a little more than if I had unlimited resources. There's something so freeing and incredible about it. The word 'empowering' has become kind of gross, but it's empowering in a sense to know that you're creating your own experience and that you're relying on yourself. And then, when it's over, there's a greater sense of satisfaction because you're like, holy shit, I just did all of that."

Zerra's comment reminded me of something Zentall told me. As we learned, the scarcity loop arose to concentrate our attention and encourage persistence in behaviors that helped us survive, like finding food. But scarcity brain evolved other elegant machinery to make the most challenging searches the most rewarding.

"If animals are hungry and have to forage for that food longer and harder than usual, they tend to value that food more than if it were easier to find, even if it's the exact same food," Zentall told me. "That extra psychological value encourages future persistence and energizes them to keep looking."

We'd chase rewards in a consequential gamble. The stronger our internal cues telling us we needed a reward—intense hunger for food, bone-chilling cold for shelter—the more psychologically fantastic the feeling when we found that food or built that fire. "I think that is translated in the modern world in which humans value things they had to work longer and harder to get," said Zentall.

It's why the first real meal we have after, say, a backpacking trip or half marathon tastes that much better. Or why that first big promotion we received after years of hard work earning low income feels much more satisfying. Or why, said Zentall, "my students who get an A in their most difficult class get much more excited about it than the A they get in the class they felt competent in. These two grades are worth the exact same for their grade point average, but they value the grades that were harder to get."

I considered this as Zerra and I bobbled through pines that opened into a wide and sloping meadow. My favorite reporting trips have always ended up being those where something goes wrong and I have to navigate problems. Like my trip to Iraq, with its sandstorms and my insane fixer. Or that unplanned bonkers descent into the Bolivian jungle.

Because of this, I've developed a motto I now use for any tribulations I face: "No problem, no story." Every story has a complication. A point where unplanned events make our life uncertain and challenging. If we shy away or pay to eliminate those, we remove challenge and gain certainty. But we also learn less about ourselves and don't become the hero of our own journey.

I also experienced this on a recent hunting trip. A company had invited me to hunt a large swath of private land with a group of other hunters. This was unlike any hunting I'd done before.

My typical hunt usually happens on public land. I'll eat crappy freeze-dried backpacking meals and sleep in the dirt. Hunger, boredom, cold, and exhaustion are part of the buy-in. Hunting public land is tougher. The odds of success significantly lower. It's best to hike deep into the wilderness, to rough it, and to be willing to stay longer than you planned.

This hunt was the opposite. We stayed in a lodge (imagine a low-grade Marriott accessible only by dirt road). Meals were big and tasty, and the rooms were warm and well appointed. During the day, when the animals bedded down, we'd return to the lodge to nap in a soft bed, check email, or watch television.

The hunting was real. But the wildlife management on the property was so optimal that we were all but guaranteed to go home with an old and large elk. The other hunters on the trip were much like those Zerra alluded to. Successful people who had a limited amount of time.

It was a good experience, but I didn't get the same emotional gravitas or sense of reward I usually get from hunting. Even from unsuccessful hunts.

It sounds strange, but sleeping in the dirt, being cold at night, eating crappy backpacking meals, and experiencing deep boredom all afternoon for something that may not pay off, although uncomfortable in the moment, makes the process more rewarding. And because I eat every ounce of the meat, I can think back to that each evening at dinner.

"There came a point where I was too extreme," Zerra told me as we covered more ground. "I would get stressed if what I owned was more than what I could easily fit in a backpack. I viewed it as a liability. There was this residual feeling of . . . if I have this stuff, I'm now responsible for it and have to take care of it. And it's stressful."

As we were moving, squirrels occasionally emerged from the underbrush or scurried up trees. They were working hard for their next meal. It turns out we humans have more in common with squirrels than we might think.

Every fall, the cold sets in and changes squirrels. It leads to a release of hormones that trigger their little squirrel scarcity brains to go into hoarder mode. The squirrels begin collecting and storing as much food as possible for winter. But these seasons can go one of two ways.

Fall hoarding is a relaxed endeavor if it's been a good summer for nuts and seeds. It's like going to the Mall of America with friends and a fat wad of cash. The squirrels gather, saying hi to their squirrel neighbors as they do. They place all their nuts and seeds in a burrow and let it be as they casually add more to it.

But the situation changes if it's been a bad summer for nuts and seeds. Scarcity makes the scene apocalyptic. The squirrels become defensive and paranoid. They think their neighbor squirrels are trying to steal their stuff. They'll fight with other squirrels over scarce nuts and seeds. And then, once they've amassed a large enough hoard, they'll prepare for more conflict. They'll stand like a bouncer at the entrance to their burrow, ready to defend the hoard.

I spoke with Stephanie Preston. She's a psychologist at the University of Michigan who has spent her career studying how animals and humans relate to their possessions. She told me that when facing a disaster that makes necessities scarce, we humans effectively become squirrels. "A good example is the pandemic," Preston told me. The pandemic was like a very bad summer.

At its onset, people panicked. Our first response was to hoard. Brawls broke out in aisles as people clamored for toilet paper, canned food, hand sanitizer, and more. And then, once we had the resources to weather the pandemic winter, we continued our squirrel-like behavior.

"This is when we saw a trend of people worrying about getting robbed of items like toilet paper or food," Preston said. "So then people went and bought guns. There were a record number of guns and ammo sold and a record number of new gun owners. This is the exact same as squirrels, who hoard, then sit at the front of their tunnel home ready to fight off competitors."

But after our initial squirrel-like pandemic freak-outs, a second phase set in. That's according to Kelly Goldsmith, the Vanderbilt researcher from chapter 4 who studies scarcity. Once people had secured basic goods, many of us began falling into a scarcity loop of mindless purchases to ease stress.

This trend got a lot of press. But what we didn't hear of quite as often is another behavior trend. Some people reacted in the opposite way. Organizing, decluttering, purging, and making everything just so also spiked during this second phase of the pandemic. Squirrels don't practice the magic art of tidying up. But we do.

Donations to the Salvation Army doubled. The amount of non-garbage items people threw out from July through September 2020 in New York City rose roughly 10 percent. The executive director of the Association of Resale Professionals told the *New York Times,* "Everyone has been overloaded with incoming merchandise. This is an experience our industry hasn't gone through before." Some secondhand stores had to rent on-site storage units to manage all the donations. The reason some purged is the same others bought. As one pandemic declutterer put it, "I now feel much calmer."

Preston has spent years studying this sort of behavior. Her work has shed light on the cycles of buying and purging that so many of us go through.

Preston explained that both overaccumulation and minimalism are "often driven by a kind of perfectionism where you want to do everything just right. There's a sort of anxiety, but it's different from overaccumulators. Overaccumulators have anxiety that they're going to make a mistake and need something, so they collect and collect. But [minimalists] have a kind of anxiety around disorder and having so much they can't escape." Whether it's keeping all sorts of odds and ends in your home for fear of possibly needing them one day or going all in on minimalism, "the behavior helps people find a sense of control," she said.

This is why blindly aiming for less doesn't work. Minimalism looks good in photos we share online, but it doesn't solve the underlying problem that we think it will. As with addiction, a better question is to figure out what's bubbling under the surface. Why do we want less in the first place?

"Everyone is a little bit stressed, because we all work so much," Preston told me. "Then with access to cheap and abundant goods everywhere, we see that many of us get caught in a loop of using stuff to assuage how we feel." It could be buying a lot of it, a bit of it, piling it, organizing it, or minimizing it. "But you get caught in a cycle that ultimately makes it worse. It ends up hurting your quality of life," said Preston.

Zerra says she now feels as if she's struck the delicate balance of having enough. "I'm in a pretty good place right now," she said. "I don't have too much, that's for sure. But I also don't feel like I'm depriving myself. Everything I own has a purpose and I appreciate it."

I considered Zerra's approach as we hiked. "I feel like you don't own stuff," I said. "It's more like you own 'gear.' Every item you have serves a higher purpose, sometimes multiple purposes, and helps you do the things that make you feel alive, like shed hunting."

She thought about it. "Yeah," she said. "I hadn't thought about it like that. But I do mostly own gear rather than stuff."

By our final afternoon, I was tiring of shed hunting. We hadn't found a thing. As casinos learned from their early slot machines, too many losses in a row leads people to quit playing the game.

But then we were headed down a wallow of dense pines and thick dead bushes. Zerra felt as if the area were "prime for elk to die."

"Look," she said. Two antlers rose from the ground fifteen yards ahead. Their off-white color and curves stood out from the brown straight twigs rising around them. She took long steps toward them, pushing back thick brush.

Zerra stood over the antlers. Then she clenched her fists and pumped her elbows, like a World Cup soccer player who scored a decisive goal. It was a full bull elk skull with both antlers intact. Jackpot.

She grabbed the antlers by their brow tines, which are the antlers that run over the elk's eyes. The skull rested on her stomach, and the white and dark main beams of the antlers hung four feet out into the air horizontally.

Her smile was childlike. Enamored and joyful. "It's beautiful," she whispered. "So exciting."

Our effort, calculation, and tenacity rendered something stunning.

Now that I understood shed hunting, I was converted. "We still have a few hours of light left," I told Zerra. "Let's keep going."

Two hours later we were in an opposite canyon. It was overgrown like all the canyons there. I had to bend every which way to shove myself through dense wooden thickets. Zerra spotted something off-white in a clearing.

A sheep skull. The ultimate find. She ran to it and I tailed closely. But as we got within a couple yards of it, she looked back at me. "Oh, no," she said. "No. It's old. The horns are rotten. That hurts. That really hurts." The ultimate find was actually the ultimate near miss. Like four of five jackpot symbols lining up and the fifth just being a bit too high to complete the big win.

The final night, we were sitting around the fire. A cold breeze was swirling the fire's piney smoke, occasionally pushing it into our faces.

"Do you mind if I ask you a somewhat strange question?" I said. "When you're searching for antlers, what does it feel like as you're searching?"

Zerra stared into the flames. "I'm 100 percent in the moment. I'm not thinking about anything but what I'm doing," she said. "I'm not thinking about what might exist in the future and what happened in the past. I'm just 100 percent there. And it's so freeing."

She then had to address the obvious. "I'm also very aware that what is getting me there, in that moment, is literally finding these discarded growths off an animal's head. Like, it's kind of weird. Actually very weird."

She continued: "But it's like suspense is building and building as I search. And searching is so challenging in the moment. But the real joy is five minutes before I find an antler, when I'm coming into an area and thinking, 'This area looks really good.' It's so unpredictable. I might think I'm in a good area and find nothing. But I'm chasing a constant feeling of hope. If I was able to just walk out there and find an antler or skull every time, it wouldn't be fun. It would lose its purpose.

"And when I find antlers or a skull, it's so rewarding. And it gives me a marker of success and an artificial purpose for being there. Being out here isn't about finding antlers. I give most away. Rather, it's this fun, totally engaged moment of being *in it*."

She paused and looked at me. I'd been nodding the whole time. "It's like a different part of my brain switches on. And what that does for my mental health, getting to exist in the moment in these searches, that is what it's all about. I realize it's weird."

What Zerra is doing is more like ultra-sanity.

I described the scarcity loop to her. I told her about the work of Thomas Zentall and how the loop is so powerful because it evolved to keep us alive. I explained how Si Redd and the casino industry leveraged it and how it's now being used and sharpened by many of the technologies that suck us in.

"Ah, so it's almost like the wilderness is my casino and shed hunting is my slot machine," she said.

Zerra uses the loop in a way that gives her purpose. It leads her to radical presence, time in nature, exercise, and improved mental health. She's turned the scarcity loop into a powerful positive habit-building loop. An abundance loop.

As we hiked out of the wilderness the next day, I thought back to Zerra and me packing for the trip. I'd done surprisingly well following her guidance.

The experience led me to a rule to guide my future purchases. I landed on "gear, not stuff." Stuff is a possession for the sake of it. Stuff adds to a collection of items we already have. We often use stuff to fill an emotional impulse or advertise to society that we're a certain type of person. Or it solves a perceived problem we could have solved better with a bit of creativity.

Gear, on the other hand, has a clear purpose of helping us achieve a higher purpose.

This approach hits all three ways we can get out of a scarcity loop. It takes away the opportunity, unpredictable rewards, and quick repetition of mindless buying. The opportunity an item provides shifts to something more meaningful. The unpredictable rewards become what the gear allows us to accomplish and experience. And the pause of thinking "gear, not stuff" before a purchase reduces how much we buy.

Zerra's gear is the items she keeps in her backpack. She doesn't own too much. Nor too little. She owns gear: items of utility that allow her to fall into an abundance loop that works her body and mind and leads to deep satisfaction.

For the rest of us, "gear, not stuff" can help us get a little closer to having our own metaphorical backpack of items that lead us to live more engaged and meaningfully. Even if our backpack is a house.

Information

I first met the astronaut Mark Vande Hei in outer space. He was in low-earth orbit, looping 18,000 miles an hour 250 miles above the surface of Earth. And I was sitting in my office in Las Vegas, firmly grounded and speaking with him on a superpowered videoconference call.

Vande Hei is an interstellar endurance freak. He's been a part of five different expeditions on the International Space Station (ISS) and racked up 523 days, 8 hours, and 59 minutes in space. He has the longest period in space of any American, a 355-day stint between 2021 and 2022.

To keep astronauts upbeat and psychologically engaged during long stretches in space, NASA will connect with anyone an astronaut feels like speaking with. Vande Hei enjoyed listening to an audio version of my last book, *The Comfort Crisis,* while in space and was curious to chat with me. So NASA made it happen.

The NASA folks in Houston arranged for me to speak to Vande Hei at 8:05 a.m. I logged on to my computer fifteen minutes early and poked around the internet as I waited. Skeptically, might I add. My neighborhood, surrounded by jagged desert peaks, hardly gets enough cell service for me to order a pizza from a guy three miles down the street. And here we were hoping to do a live videoconference with a person inside what is effectively a man-made asteroid hurtling around the planet.

"Hi." I heard a voice come through the speakers and clicked back to the video feed app. "Thanks for taking the time to talk to me."

And there Vande Hei was. I knew this wasn't some long con because the man was floating. For real. He was dressed casually in tan pants and a tucked-in blue T-shirt. The live video feed was surprisingly crisp; your tax dollars hard at work.

He opened the call by giving me a quick tour of the station.

The inside of the ISS looked as if a techno hoarder who has a thing for the color white designed it. Every surface was coated in an array of wires and switches and buttons and screens and dials and tubes and panels. Without gravity, there's no such thing as a ceiling or floor. All is wall. There's no real up, down, or sideways.

He showed me the kitchen, which was arguably the cleanest and most organized kitchen in the universe. Dehydrated food contained in individual sleeve-shaped packages lined a metal drawer like tax folders. Next, he led me to the room that holds the space suits. Then the gym. It had strange contraptions attached to every wall. The Russian cosmonaut Pyotr Dubrov was horizontal in the camera frame, standing on the wall and doing squats on the gym's Advanced Resistive Exercise Device. It uses vacuum-sealed cylinders to produce up to six hundred pounds of resistance.

In our ninety minutes of conversation, Vande Hei would orbit Earth exactly 1.29 times. Astronauts loop Earth so fast that they see the sun rise and set sixteen times daily.

"Our primary activity up here is science," Vande Hei told me. "We're here to get information." ISS astronauts, for example, recently sequenced DNA on the station, studied the effect of gravity on the optic nerve, looked at how fire behaves in a zero-gravity environment, and much more. In space, systems we ignore and take for granted on Earth behave differently. Vande Hei explained that we'll need to decipher those differences to successfully go deeper into space and understand ourselves and our future.

"Every day, there's something new I learn," he said. "The whole idea behind science is doing and learning new things. The desire and opportunity to explore and get a different perspective drew me to being an astronaut. Exploration is about new experiences and seeking answers to bigger questions."

He admitted that this exploration isn't always easy. Astronauts must sacrifice so much to break new ground in human understanding. There's the psychological toll of spending a year away from family and enclosed in one large, optic white, loud, and uncomfortable ship that's the size of a Boeing 747. But also a physical toll. "I'll probably lose 8 percent of my bone density on this mission," Vande Hei said. Meaning he'd return to Earth with the brittle bones of a centenarian. He'll have to limit his running for a while, and it'll take him a year or two to regain that bone density. One of the rate-limiting steps of getting to Mars—a mission that would take at least three years—is figuring out ways to maintain the human body.

But Vande Hei said the hardships are entirely worth it. "Let me show you something," he said.

He let go of his computer and flicked it down a long hallway. The bright white walls of the ISS darkened. The screen stabilized and Vande Hei was back on-screen, holding the computer facing him.

"Here's Earth," he said. He turned the camera 180 degrees.

A quarter circle hung in an infinite void. A thin blue line separated the quarter circle from the black void.

"That varnish of blue is the atmosphere," said Vande Hei. It is the only thing between us and a hot death. "That's where we live. We don't live in the earth. We live in the atmosphere." My home, your home, our home.

"Right now, we're going over Canada, Michigan, and the Great Lakes," he said. The top of the mitt of Michigan and the Upper Peninsula were mostly white from snow, which was fractaled by a bit of brown. The shores of the Great Lakes were an aquamarine that smoked out into a body of steel-blue water. Westward was dark.

I've grappled with whether life came from nothing, means nothing, and proceeds nowhere. And, well, let's just say seeing Earth from this perspective made me consider, quite deeply, that there's some higher order to this all.

Lake St. Clair was lighting up a baby blue. The Detroit River emerged from it, bisecting its namesake city and running into Lake Erie. If I can give us any hope for our future, a reason to be a good and helpful human, it is this: even Detroit looked beautiful from up there.

Vande Hei must have known this vantage point can affect people. He was silent for a moment to let me take it in. Eventually, his voice returned.

"After my first mission on the station, I would have said that you can't see the stars very well from the space station, because we have too many lights on inside," he said, with the camera still pointed toward Earth. "But then I started coming up here a half hour before everyone else would wake up. And I'd shut off the lights and just sit with my eyes open and sort of meditate, not trying to think about much, but rather just trying to soak it in. It was challenging at first, but ultimately amazing. My eyes would adjust. Space actually isn't the inky black that I described to people after my first mission. It seemed to me that the stars we normally see as individual stars weren't really that way. I could start picking up the faint stars super far away. There were so many of them that it seemed more like shades of gray and white and black. It was more of a texture, a net. And that changed how I think about perception, the limits of human knowledge, and how we have so much more to explore."

Astronauts follow a long line of explorers. Scientists believe that life has existed on our planet for 3.75 billion years. Life started as a cell in the ocean. But thanks in part to exploration, that cell metamorphosed over time. It evolved into all the bacteria, bugs, birds, fish, and animals that ever lived. *Homo sapiens,* our species, emerged from this sequence between 200,000 and 300,000 years ago.

This exploration was always challenging. But it was required for animals to evolve and improve their lives. Humans are the greatest example of this story.

The unknown calls us loudly to make itself known—to find opportunity and progress. But now abundance and the scarcity loop are combining to alter our drive for information and exploration in ways affecting us.

No other species explores like humans. Most other animals commit to a range they're built for and stay there from generation to generation. Even when other species migrate or disperse, they do so within predictable areas. For example, caribou don't migrate to Miami. Penguins don't march to Minnesota. Steelhead trout don't swim to Seaside Heights, New Jersey.

Even the animals that have spread widely don't explore like us. For example, foxes are all over Earth. But they've changed as they spread. Hence there are twenty-three different types of foxes that each have their own environmental niche. Bengal foxes, arctic foxes, red foxes, island foxes, pampas foxes, and so on.

Meanwhile, there's now only one kind of human, *Homo sapiens.* And we can go or live anywhere. We've summited Earth's highest peaks and dived to the lowest points of its oceans. We've built more than fifty active research stations in the North Pole, set up outposts in the thickest reaches of the Amazon, and constructed modern cities in deserts that should be all but lifeless.

We cherish our ability to do this. Many of our greatest tales are those of an everyday person taking on an epic quest, encountering the unknown, and having the experience transport them into a higher state of consciousness. We call this the hero or heroine's journey. It exists in the stories of all cultures. The *Odyssey, Arabian Nights, Moby-Dick, Heart of Darkness,* the *Epic of Gilgamesh, Alice's Adventures in Wonderland, The Lord of the Rings,* and on and on.

Our path to becoming great explorers took more than three billion years. To understand it, I called Neil Shubin.

Shubin is an evolutionary biologist at the University of Chicago. At work he studies life on Earth and how it came to be. On the weekends, he's a midwestern, dad-of-the-year type who runs a lot of sidelines at youth soccer games.

This is why it's odd to picture Shubin in one of Earth's most savage places: Ellesmere Island, an uninhabitable mass of tundra and ice in the Canadian Arctic. It's part of the Arctic Archipelago, a cluster of islands off northern Canada that are one of the northernmost points of land on the planet.

"The island is a high arctic desert," Shubin told me. "Imagine the American Southwest—red sandstone and buttes and mesas—but it's in a polar landscape with glaciers, ice, tundra, and polar bears."

In 2004, a helicopter dumped Shubin and his small team of scientists on the southwest corner of the island. They were there searching for something that had been dead for hundreds of millions of years. Shubin had spent every July for the previous six years on Ellesmere. He'd burned millions of dollars in research grant money on these searches. He hadn't found a thing.

But Shubin was obsessed with a fascinating feature of the human arm. Our arms are built like so: one bone attaches to our torso (our humerus), two bones attach to that one bone (our radius and ulna), a little blob of bones attaches to those two bones (our wrist), and then our fingers attach to the blobs. This is the *exact same* architecture of the arms of lizards, bears, dogs, cats, lions, tigers, bears . . . you name it. Every land animal has that one-bone-to-two-bones-to-blob-to-fingers structure.

The same goes for our legs, lungs, livers, eyes, ears, mouths, and so on. We look very different, but all land animals have similar architecture. A sort of blueprint for their bodies. But where did that blueprint come from? It had to come from something.

Shubin was searching Ellesmere Island for that something. By examining older and younger fossils, he had deduced that this something must have emerged around 375 million years ago. But digging randomly for fossils is like throwing a dart at a globe-sized dartboard. Ellesmere had 375-million-year-old sedimentary rocks, which are the type of rocks that can contain fossils. And those 375-million-year-old sedimentary rocks were exposed and not covered by grass or strip malls.

And if the exposed sedimentary rocks of Ellesmere held that theoretical something, that blueprint, and Shubin could find it, he'd have made one of the greatest discoveries of all time. It was a scientific swing-for-the-fences endeavor.

But 2004 was a "do-or-die situation," Shubin told me. "We were running out of money." Funders, like gamblers, stop paying if they string together too many losses in a row. And all signs were pointing to another bust.

On one of the final days on the island, Shubin was at the bottom of a quarry. The place was like the surface of Mars, with piles of blaze-red rock covered in kaleidoscopic blue and white ice. He was doing what he'd done nearly every July day for the past six years: knocking around chunks of rock and ice, searching for that theoretical something, and fretting about another failed expedition and falling into the large and forgotten pool of scientists who couldn't back their big notions.

While white knuckling his hammer, he noticed a chunk of ice covering some rocks. It was like every other chunk of ice and rock he'd been hammering. But he hammered it anyway and "saw something I'll never forget," he told me.

Scales. He'd found plenty of scaly fossils on the island before. But these scales were different. Instead of being big and smooth, they were all laid out like small and rough clay shingles atop an adobe roof.

His heart started cranking. He hammered more. More scales appeared. More hammering. And then the big reveal: a set of terrifying jaws.

The history of life before animals appeared on land essentially goes like this: Roughly 13.8 billion years ago, the big bang happened and created outer space. Over time, galaxies began forming. Earth formed about 4.8 billion years ago. Our planet was an uninhabitable and constantly changing solid ball. But soon an atmosphere enveloped the ball. Water then pooled on the ball's rocky surface and filled its low areas to create oceans and lakes.

Life formed around 3.75 billion years ago. It appeared in the ocean as a single microscopic cell.

Eventually, around 800 million years ago, some of these cells grouped together. This allowed them to find more resources and safety in numbers. These groups of cells eventually broke into smaller groups within their larger group and took specific jobs. For example, some of these groups sensed the location of food. Another group helped process that food. Which means that these small groups became organs within a larger body: brains, digestive tracts, and everything else. This is how we got the first animals.

By 580 million years ago, all kinds of freak aquatic animals were filling the oceans. They had bodies shaped like sponges, ribbons, and spiny magic carpets. These sea creatures eventually began to develop shells and spikes for protection and gnarly heads and tails so they could move fast and hit hard to kill and eat others.

Then, about 400 million years ago, Shubin told me, "you had a place for creatures to live on land. Plants and insects had developed root systems and soils. So on land there's this whole new ecosystem to explore, but most of life was still evolving in the waters."

And these waters were now a full-on "fish-eat-fish environment," said Shubin. "There were small fish and big fish, but they were pretty much all predators. And I'm talking *big* fish—fifteen feet long with giant teeth the size of railroad spikes."

Teeth just like the ones Shubin found himself staring down at in that quarry on Ellesmere Island.

Shubin and his team carefully excavated the fossilized creature and took it to their lab. The fish looked like a cross between an eel, trout, and alligator. It was nearly ten feet long. And it was a true underdog for its time. Resources were scarce and life was dangerous for this fish. The bigger, faster fish were trying to eat either the fish or its food. But this fish had a distinctive asset: strange fins.

These fins were uniquely constructed—one thick bone attached to two bones attached to a blob attached to some fingerlike things. This allowed the fish to prop itself up and waddle along the ocean floor. The way a crocodile walks.

One day this fish used those special fins to waddle out of the water and onto land. It was the greatest and most consequential exploratory expedition in the history of our planet. This fish was the original Hernando de Soto or Ferdinand Magellan—heading into the unknown to gain information that might bring it a better life. Land, this creature would discover, offered more. More food. More resources. More safety.

In doing this, the fish brought a blueprint to land.

Shubin named the fish Tiktaalik, which is Inuit for "large freshwater fish." It's now well accepted in scientific circles that Tiktaalik's arms, lungs, neck, teeth, nose, eyes, ears, and so on provided the blueprints for our own arms, lungs, necks, teeth, noses, eyes, ears, and so on. And also for all other animals that have lived on Earth—from velociraptors and puffins to lions, tigers, and bears. Tiktaalik gave life on Earth a rather straightforward genetic recipe. The recipe has simply evolved for each creature in ways that allow it to get more and live on. If it doesn't? Extinction.

"And this is not at all speculative," said Shubin. "These are beautiful connections that have multiple lines of evidence over a century and a half of scientific research."

In April 2006, Shubin published his research on this fish and what it can tell us about ourselves. The study appeared on the cover of the scientific journal *Nature*.

Beyond bringing a blueprint for all other animal bodies, this fish perhaps carried something of a drive to explore. A permanent itch for information that could improve its life and a willingness to keep on scratching it until it paid off. Tiktaalik "had the right inventions at the time for it to explore land," Shubin told me. "So it stepped out of water because of the opportunity it hoped to find."

Opportunity—> Unpredictable Rewards—> Quick Repeatability

Scientists once believed hunger, thirst, sex, or some other survival need drove everything humans did. But Harvard psychologists in the 1950s discovered that humans and many other animals are driven to explore.

The research had been accumulating for years. In the early twentieth century, Ivan Pavlov, the psychologist who famously trained dogs, noticed that his dogs had what he termed an "investigatory reflex." That's a bit jargony, so Pavlov often just called it the "what is it reflex." When he put a dog in a new place, the dog immediately ran about sniffing—exploring for information to answer, "What is it?"

He also saw the "what is it" reflex in himself, his colleagues, his friends, his family—everyone. Then scientists began discovering that exploration is critical to our development as humans. Newborns, for example, scan and visually explore new scenes for far longer than old ones. This helps brain development. Other research shows that infants and toddlers who are allowed to explore the world develop better and faster than those who are helicoptered and kept mostly in the same place. The explorer kids gain language and physical skills faster, build stronger immune systems, and better understand the world. They even sleep better.

Pavlov explained that the reflex "brings about the immediate response in man and animals" to explore new things and places.

Soon after, in 1925, scientists built on Pavlov's ideas about exploration. They discovered that rats would cross an electrified grid that shocked them just for the privilege of entering a new territory.

Other studies found that to explore, monkeys will put themselves through four hellish, frustrating tests—the monkey equivalent of taking the bar exam.

What's critical is that none of these lab animals were hungry, horny, thirsty, cold, or in danger. In all cases, these animals explored when all their needs like food, sex, water, and safety were taken care of. They explored just to explore.

So by the 1950s, the Harvard scientists felt confident writing, "One of the most obvious features of animal behavior is the tendency to explore the environment. Cats are reputedly killed by curiosity, dogs characteristically make a thorough search of their surroundings, and monkeys and chimpanzees have always impressed observers as being ceaseless investigators." And so, the researchers proclaimed, "exploration should be listed as an independent primary drive."

Although humans share this drive with other animals, our drive to explore is stronger—taking us farther and deeper into the unknown. Just as Tiktaalik had special fins that allowed it to waddle into the great beyond of land, our human bodies and brains have features that make us even more ceaseless investigators.

Scarcity shaped us into epic explorers. Recall from chapter 4 how Earth experienced a great cooling period between 9.3 and 6.5 million years ago. At that time, our ancestors were much more apelike than we are now. They lived in the jungles of Africa and survived by eating fruits. But as the world became colder, our jungles began shrinking. The outskirts of the jungle morphed into much drier woodlands. Fruit trees thinned out. Fruit became harder to find and available only during certain seasons.

In this new environment of scarcity, the best explorers thrived. The apes that had bodily quirks that helped them cover more ground and stand taller to reach for fruits were able to get more and better food. They in turn survived and spread their genes. This kick-started the evolution of apes into humans.

Scarcity is the mother of movement, invention, ideas, and breaking new ground on the map and in the mind.

Over time, natural selection did its thing. We became, as the Grateful Dead put it in their song "The Wheel," "bound to cover just a little more ground." Our ancestors developed features that allowed us to stand tall and walk on two feet instead of moving via the four-limbed "knuckle walking" like most other primates. We got arches in our feet, large knee joints, knees angled under our hips, a large hip joint that faces sideways, a tall narrow waist, a long spine, and more.

Walking upright on two legs helped us travel greater distances and freed up our hands. This gave us two unique advantages.

First, we could reach more scarce food in trees and then carry that food back to camp. Humans are the only animal that can carry weight across long distances of their own volition. We could then also carry tools and other supplies into the unknown. These tools helped us survive in unfamiliar places.

Second, walking made us far more efficient at covering ground. For example, it costs other apes four times as much energy to travel the same distances as us humans.

Still today other apes like chimps walk an average of only a mile or two a day, while modern hunter-gatherers walk an average of eight. But sometimes our ancestors would travel more than twenty miles in a single day during big hunts or explorations. Today, ultramarathoners can cover a hundred miles on foot in less than twenty-four hours.

Researchers from Yale say our extensive foot travels created an epic exploratory feedback loop. The more we explored, the more resources we could get, especially from food. The more resources we could get, the more we could fuel the development of our amazing brains. The more our amazing brains developed, the more we could figure out how to explore new territories.

This might be why still today walking while paying open attention to the world can enhance creativity, concentration, and understanding. A recent study in *Psychological Research* discovered that a group who walked freely with their awareness on the open world scored significantly higher in a creativity and idea test than people who walked while focusing on their phones. Another study found that children who walked for twenty minutes improved their concentration and ability to understand complex information. The scientists wrote that unfocused walks could "support cognitive health and may be necessary for effective [brain] functioning across the lifespan." Enhanced focus and creativity while walking might have helped us note important landmarks, dream up ways to go farther, and survive while exploring.

Many animals have features that help them thrive in a specific location. Like, for example, the white fur of an arctic fox. But our brains allowed us to become generalists. We can think creatively and envision unique uses for our tools. While other animals may see a fallen tree as an obstacle on the ground, humans can picture that fallen tree being dug out with a rock, placed in water, and used as a canoe. Once our canoe is in the water, we'll still keep thinking abstractly to go farther. We dream up ideas like attaching a sail and a rudder.

As the Nobelist and director of the Max Planck Institute for Evolutionary Anthropology, Svante Pääbo, told *National Geographic*, "No other mammal moves around like we do. We jump borders. We push into new territory even when we have resources where we are. Other animals don't do this. Other [ancient human species] either. Neanderthals were around hundreds of thousands of years, but they never spread around the world. In just 50,000 years we covered everything. There's a kind of madness to it. Sailing out into the ocean, you have no idea what's on the other side. And now we go to Mars. We never stop. Why?"

Why, indeed? What are we searching for? Scientists now believe that our drive to explore is ultimately a search for information.

Dr. Judson Brewer, a psychologist at Brown University Medical School, told me that as humans evolved, our scarcity brain developed a craving for information. Especially information that improved our life and increased our odds of survival.

Having more information made us far less likely to die at any moment. It paid to figure out where our next meal was coming from. Or to discover new areas for food and other resources. Or to know whether a storm might be rolling in, understand other people's motivations, or predict the future. The more information we could get, the more likely we'd acquire food, sex, stuff, status, and more and avoid the situations that might kill us.

It's why Tommy Blanchard, a researcher at the Harvard Computational Cognitive Neuroscience Lab, referred to humans as "informavores: creatures that search for and digest information, just like carnivores hunt and eat meat."

Our searches into the unknown to seek information rode the scarcity loop. They were powered by longing and wonder for greener grass and required deep engagement of our mind and body. We had to be all in. We'd set out from home into the abyss. Knowing nothing. What may come was unpredictable. Over the hill could be a large group of animals that we could eat—or animals that might want to eat us.

And we'd continue exploring—repeating the cycle—until we found that greener grass and were rewarded for it. That is, until we realized there was probably even greener grass elsewhere, thrusting us right back into the loop of opportunity and unpredictable rewards.

Scientists have even discovered a gene, called DRD4-7R, that is linked to exploration and a willingness to take risks. Scientists have nicknamed it the "wanderlust gene."

But the gene isn't linked only to exploration. The science writer David Dobbs explained that the gene seems to "make people more likely to take risks; explore new places, ideas, foods, relationships, drugs, or sexual opportunities; and generally embrace movement, change, and adventure."

NASA once released an official explanation that stated, "Humans are driven to explore the unknown, discover new worlds, push the boundaries of our scientific and technical limits, and then push further. The intangible desire to explore and challenge the boundaries of what we know and where we have been has provided benefits to our society for centuries. . . . Curiosity and exploration are vital to the human spirit."

And maybe even our survival. The philosopher Frank White, in his influential book *The Overview Effect*, wrote, "Space exploration may be a key to human survival and evolution, and perhaps even more than that. . . . [W]e are not simply reaching out into space to use extraterrestrial resources and create opportunities here on Earth. Rather, we are laying the foundations for a series of new civilizations that are the next logical steps in the evolution of human society."

Vande Hei and I were still looking out at the half dome of Earth when he indicated he agreed with White's hope for space exploration. But with a warning.

"Ideally one day we'll be successful in space exploration long enough that we can get farther and farther away from Earth and reach and live on other places," he said. "But even if we find decent environments, I can't imagine they'll be as comfortable for us as Earth, because we're completely adapted to living there. So we have to perceive that Earth is our primary space to exist; otherwise we're at risk of throwing it away."

Earth was still scrolling slowly across the video feed like a screen saver as Vande Hei told me to look to the northernmost reaches of what we could see.

"I grew up in Minnesota," he said. "I've had a lot of moments here looking outside at the earth and its wide open spaces. I'll look at all the lakes and waterways of Minnesota and Wisconsin, or the coastlines of Labrador in Newfoundland, or even the vast interior of central Asia. They're all so empty. From up here you see that Earth is still mostly vast spaces that don't really have any people. I now have a lot more curiosity about the earth than I did before. When I get back, I want to get outside more and go into new areas."

Most of us today aren't exploring in the traditional sense. Going into space like Vande Hei. Or spending half the year breaking new ground in the wilderness like Zerra. But we all still explore.

We're all informavores. The drive to know the unknown still exists in us all. It powers the unpredictability of the scarcity loop. It's that deep, angsty discomfort we feel as we wait for information about an outcome. That could be knowing the falling of slot machine reels, a flood or drought of likes, the shifting of a Robinhood stock, or waiting for a rightward swipe.

But in our past, information was scarce, limited to what our senses could take in, in person, in the present moment. We seem to have always believed that more information could improve our lives. But we also lived in a world of informational scarcity. If we wanted new information, we had to physically "go there" to get it.

As humans evolved, all the world was empty and untamed. Seventy percent of that land was livable for humans. The remaining 30 percent of the land was what scientists call "barren land." Glaciers, dry salt flats, beaches, dunes, rocks, and so on. These figures mostly hold today but may be shifting due to climate change.

We might be surprised to learn that still today the world is mostly vast and unpopulated spaces. Our urban areas take up just a sliver of Earth. Cities, towns, and villages make up only 1 percent of our habitable land. Most of our habitable land, 50 percent of it, goes to agriculture.

The rest of it—millions and millions of square miles—is open for us to boldly go. In the United States alone we have 640 million acres of public land, hauntingly beautiful and wild country that is ours to explore. That's roughly equivalent to the size of six and a half Californias. But we rarely enter into wilderness. For example, in national parks only 14 percent of visitors meander beyond a paved road.

Edward Abbey wrote, "You can't see *anything* from a car; you've got to get out of the goddamned contraption and walk, better yet crawl, on hands and knees, over the sandstone and through the thornbush and cactus. When traces of blood begin to mark your trail you'll see something." He's a bit extreme. Still, science and thinkers have long known about the benefits of getting off the beaten path. The research suggests time in the wilderness does help us "see something." As I dug into writing *The Comfort Crisis,* I learned that nature profoundly improves our mental, physical, and even spiritual health. The wilder the nature the better.

Beyond the wild, we also have millions of blocks of small towns and big cities we can explore. Pockets of culture and a chance to learn, hear, smell, taste, and touch new ways of thinking about and being in the world with others. Even still, research shows most of us slip into a predictable routine. We take the same route to work and frequent the same neighborhoods, restaurants, and businesses.

Today our drive to explore and expand our horizons hasn't been tamed out of us. Rather, scientists at Temple University say that how we explore has changed radically over the last two decades. They wrote that just as our ancestors searched for information on the savannas, today we search for information online.

It's nice to hear that exploring Amazon for the best deal on, say, a coffeemaker is similar to our ancestors exploring the African savanna. But it's also kind of depressing.

We can now search for information and advice anytime, any place, up to the second, and spiral into an informational matrix—hoping to find greener grass. In a more controlled, comfortable, inactive, mediated setting.

Even until two hundred years ago, information was still relatively scarce. It was mostly what our senses could take in, in person, in the present moment. Roughly 15 percent of the world was literate. If we wanted new information, we still had to get it through a mind-body effort.

Humans lived like this—more present and focused on information relevant to their life in the moment—until roughly 1833. This is when a man named Benjamin Day started a newspaper. He was the first to realize that his "product" wasn't his newspaper or the news. It was his readers and their attention, which he could sell to advertisers. He understood that the more readers he could gather, the more he could charge per ad.

So he did two radical things. First, to get more readers, he made his paper six times cheaper than other papers. This move allowed more people to afford his paper. He was selling at a loss, but he planned to make his money back in ad revenue.

Second, he leaned into the type of information scarcity brain fixates on. Papers at the time covered boring topics like business. Day realized that to gather eyes and make more money off ads, it's best to run stories that leverage the unpredictability feature of the scarcity loop. Recall that our attention gravitates toward unpredictable information that may result in great reward or pain and suffering. For example, watching slot machine reels line up indicating a possible jackpot or hearing about a serial killer on the loose.

Day leveraged the latter. He ran negative stories: tales of mayhem, murder, deceit, theft, outrage, gore, scandal, and so forth.

Within a year, he had the largest paper in New York City and copycats were popping up. This began the attention-capture and information economy in which we now live. And there's been an arms race to leverage our thirst for information to capture our attention ever since. Still today, roughly 90 percent of news is negative.

A group of scholars in the U.K. wrote that when radio arrived in the early twentieth century, it "finally cracked the secret of immediate transmission of information in real time to the masses." Radio programmers learned that media could "own" people's days, feeding them constant information. Then we got TV in the 1950s. It was the ultimate information portal. In just a decade, the average person went from watching no TV to watching five daily hours of it.

Then we got the internet. This altered the source of mass information. It no longer came from some broadcasting or publishing company on high. The internet allowed anyone anywhere with a modem to be a broadcaster or publisher. To fire information into the ecosystem that anyone could read, watch, or hear.

The result: At the start of the twentieth century, humans spent no time taking in digital information. By the 2020s, the average person spent between eleven and thirteen hours of their day consuming information on-screen and through speakers. Now 40 percent of this content is "user generated." It's the YouTube and TikTok videos we watch, blogs and Reddit threads we read, and many podcasts we listen to.

Some scholars estimate that in one day we are now exposed to more information than a person in the fifteenth century encountered in their entire lifetime. Much of it leverages the scarcity loop to make us feel self-righteous, outraged, happy, sad, or correct—all so we'll see ads.

The Columbia University media scholar Tim Wu explained, "A consequence of [the advertising business model] is a total dependence on gaining and holding attention. This means that under competition, the race will naturally run to the bottom; attention will almost invariably gravitate to the more garish, lurid, outrageous alternative." This trickle of unpredictable negative information grabs us, leveraging the loop.

And this affects us. Consider, immediately after the 2013 Boston Marathon bombings, researchers from the University of California, Irvine, investigated two groups. The first group was made up of people who watched six or more hours of televised bombing coverage. The second group was people who actually ran in the 2013 Boston Marathon.

The finding: The first group, the bombing news bingers, were more likely to develop PTSD and other mental health issues. That's worth restating: people who binge-watched bombing news on TV from the comfort of home had more psychological trauma than people who were *actually* bombed.

Another example: One investigation discovered that minutes after a group of nine-year-old kids signed up for TikTok, the app was feeding them information that suggested that COVID was a genocidal conspiracy. Like, "Hello, kids! Here's a dance video, and another dance video, and did you know that COVID was created in a secret governmental lab to kill you and everyone you love? Now please enjoy another dance video."

Much of our modern information is produced or processed not by humans but by computers. The same calculations, production, and analysis of information that took humans days can now be done in milliseconds by software. This has many upsides. For example, spreadsheets that do dull work or safety systems that can stop our cars before we realize a pedestrian is in front of us. But our ability to make sense of all the information and make decisions based on it hasn't evolved in step, wrote German scientists who studied the phenomenon.

The researchers explain that information hits a rate of diminishing returns. When we know nothing, adding information helps us make better decisions. But if we continue piling on information, we hit "information overload." At this tipping point, more information usually leads to worse decisions. The more complex information we deal with, the sooner we hit the tipping point.

But as informavores we have no clue when we've hit this tipping point. Scarcity brain still craves more information because it evolved in a world where information was scarce and more was better.

Psychologists gave a good rule of thumb for making decisions in a sea of information. It's similar to the rule we can use to determine if we should keep or discard an item. Make everyday decisions within sixty seconds. After that, analyzing more and more information only wastes time and doesn't steer us into significantly better outcomes.

In the past, we had to either accept not knowing or enter the unknown to find out. We were constantly weighing trade-offs, determining whether scratching our information itch was worth it. But now it takes no effort to scratch, and we scratch so much that we bleed.

Today, the information ecosystem expands to any question we can dream up. The internet contains billions of answers. There are millions of hits for diet and exercise, productivity and personal relationships, what products we should buy and what stocks we should invest in, where we should go to dinner or what movie we should watch.

It extends to our most profound questions, like how we can avoid death and why life exists. But it also oozes down to the most mundane details of first-world life. For example, the *New York Times* ran a story in 2019 titled "Should You Take Your Shoes Off at Home?" And it was eight hundred words of experts and scientific research weighing the pros and cons and finer points of whether your shoes should remain on your feet when you are inside the building in which you live.

This story isn't unique. I recently scanned the Smarter Living section of the *New York Times*. In just a few weeks they'd published stories that investigated topics like "how to shop for a cake stand," "how to walk a dog in the rain," "how to get the most out of your dishwasher," "how to share a bed," "how to choose the right size storage container," and "how to wake up on time."

I had always assumed that the answers to those questions are "buy the cake stand you like," "put on a jacket and walk the dog," "use it to wash dishes," "sleep on one side of the bed," "choose the storage container that fits your stuff," and "use an alarm clock." But apparently there are deeply obsessive-compulsive approaches to all of these topics. Each story spent hundreds of words and cited all sorts of experts and scientific research on the pros and cons of different cake stands, tactics for dog walking, dishwasher use, bed sharing, stuff storing, and opening your eyes in the morning. There was even a deep investigation of a "stress-free way to hang art." That story quoted five different "art-hanging experts." Five of them, all helping you solve a problem you probably weren't stressing about. Until now . . .

It's not just that we, personally, can't know everything. It's also that we now have so many experts with competing views that we can't even find the right experts. We must pick which expert to trust from a field of hundreds, even thousands. Many of them disagree entirely. For example, licensed doctors, dietitians, psychiatrists, and exercise physiologists might tell us completely different ways to eat, think, and move to be healthy.

This is the philosopher Elijah Millgram's "great endarkenment" again. He explains that today's world contains thousands upon thousands of experts in niche academic fields and work specialties (like art hanging). These "experts" have thrust so much information into the world that it's impossible to understand the deeper nuances of most topics. Roughly three million scientific journal articles are published each year, and the number grows 5 percent annually.

So how should we balance our informavore brains in our modern abyss of information? I was slapped in the face by "the great endarkenment," how easy it is to get information, and how that can give us an incomplete picture of reality when I was an intern at *Esquire* magazine in my early twenties. This was around 2010. A senior editor pulled me into the conference room to teach me how to find better information.

That editor had given me a strange reporting assignment. I was to find out how much money the pope makes. So two other interns and I dove in on the project. We searched around the internet and read a few stories. I even interviewed a Catholic academic and historian at a university. That historian hemmed and hawed and gave me their best estimate. Then we emailed our editor the research file so he could pass it off to a writer.

We got an immediate email response: "Meet me in the conference room in five minutes."

It was the end of the day and our editor was sitting at the head of a long table. We could see down the barrel of Eighth Avenue as we entered the glass-walled conference room in midtown Manhattan.

We all took a chair. Our editor took far too long to speak. Then he sighed heavily. "Guys, no," he said. "No. No. No. If you want to know how much the pope makes, you call the fucking Vatican."

"Call the fucking Vatican." +39-347-800-9066. In the years since, the line became shorthand in my mind for how I should exercise my human informavore muscle. It's a rule of thumb we should all consider anytime we want to understand something deeply—constantly questioning where our information is coming from and, whenever possible, going to the source.

The philosopher Thi Nguyen explained that there's a difference between knowledge and understanding. Knowledge is possessing facts. Understanding is different.

"First," wrote Nguyen in a 2021 paper, "when we understand something, we not only possess a lot of independent facts, but we see how those facts connect.... Second, when we understand something, we possess some internal model or account of it which we can use to make predictions, conduct further investigations, and categorize new phenomena."

Understanding is most likely to land when we work a bit harder to get our knowledge straight from the source. It comes through calling the Vatican, as it were. This requires more effort and deeper exploration. A willingness to go there, pick up the phone, or at least read a primary source, like a study. But it leads to deeper, more accurate understanding.

Want to know what something looks or feels like? See it or experience it. Curious about what someone believes? Ask them. More is revealed in person in the present moment.

We crave information, but we'd prefer it to be easy to get. Picking up the phone or meeting someone in person is more uncertain, unpredictable, and uncontrollable than being behind a screen and reading what someone else already interpreted.

This falls in line with a phenomenon researchers call "online brain." A team of scientists from some of the world's most elite institutions like Harvard Medical School, King's College London, and Oxford recently gathered to study online brain. They say that the internet has altered our minds in three ways.

First, it hurt our ability to focus. Not shocking. Our work and learning devices are the same as our social devices are the same as our fantasy football tracker are the same as our TV are the same as our . . . you get the point. Apps that leverage the scarcity loop fracture our attention and kill the focus required for deep understanding.

Researchers at Stanford found that people switch between tasks on their laptop as frequently as every nineteen seconds. More than half of college students admit to being unable to study for ten minutes without checking their phone or opening a screen for entertainment.

The second effect of online brain is that we've off-loaded some of our memory to the cloud. There are clear upsides—like the fact that we now have an encyclopedia in our pockets. But the downside is that this can make it harder for us to make connections between seemingly disparate pieces of information. It's as if we don't have access to the pieces we need to fill the puzzle. Instead of the puzzle pieces all being out on the table, some are in one room, others are in another.

Research backs this idea. One study asked two groups of people to find information. The first group could use the internet, while the second used printed encyclopedias. The internet group—not shockingly—found the information quicker. But soon after the task, they scored significantly worse on their ability to recall the information accurately compared with those who used the printed encyclopedias. The study suggests that if we want to better remember information, searching for it more labor intensively, like finding the right book, then finding the right section in the book, can be advantageous. Just as slow food has advantages over fast food, slow information is often better than fast information.

Third, the scientists say, the internet is changing social interaction. Our brain seems to respond to social interactions online and in person similarly. But some studies suggest that the rise of the internet led social anxiety among young people to triple since 2008. The researchers point to "24/7 direct and indirect media." This might be why many of my university journalism students struggle to reach out to and speak with sources. And, of course, online comment sections seem to be HQ for heinous human behaviors.

The Stanford researchers reported that the internet and its mind-bending tendencies extend everywhere: to our daily social, professional, intellectual, and private lives.

I mentioned this phenomenon when I spoke with Vande Hei. He said, "In the last five years, when I give public presentations, I've had more people ask me if the earth is flat. I actually had a streak where every single time I talked to a school, a person would ask about the earth being flat. I didn't know how to respond to that. It's made me really think about how one of the things science tries to do is come up with a theory and try to disprove it. So we go out of our way to look for information that would contradict what we think reality is. That's how we can be confident that we've got a workable theory. If we can't find any way to disprove it, then we get more and more comfortable with it, but we still call it a theory because we could learn more down the road."

Vande Hei continued as we looked at Earth, which was not flat. "So what I think is happening is that we have access to so much information that could be factual or could be inaccurate. When we have a question, we tend to go looking online for information that only reinforces the ideas we already have. You can easily find that reinforcing information. And you can get more entrenched in any idea."

Nguyen explained that discovering what we think is the right information feels good. The philosopher Alison Gopnik went as far as calling it an "intellectual orgasm." That's a line that only someone who thinks for a living could utter, but she was getting to something important. The "aha" moment feels good, like a jackpot.

Confusion, on the other hand, is an uncomfortable cue signaling us to think more and search for more information. When we find what we consider the right information, our discomfort shifts to comfort. "Aha!" we exclaim. The feeling of clarity this brings not only is comforting and rewarding but also tells us we don't have to find more information or think more. Once we have an "aha" moment, we don't need to have another. Think of this like hunger versus feeling full. Lacking information is like having an empty stomach. Finding information is like the full and happy state of finishing a hamburger.

This "aha" feeling of clarity often comes with truly understanding something, but not always. We can feel clear on a topic and be kind of wrong or even wildly wrong about it. We've all experienced this more than we'd like to admit. Remember from Chapter 7 that people who felt 99 percent certain were wrong 40 percent of the time.

There are still plenty of places where it's easy to see where we're wrong. For example, we can see that our information on how to build a bridge, cook a steak, or style our hair was wrong if the bridge collapses, if the steak has the consistency of leather, or if our hairdo makes us look like a wacko. But most of our decisions are murky.

"We are finite beings with limited cognitive resources," wrote Nguyen in a paper. "In daily life, we need to figure out what to do: where to spend our money, who to vote for, which candidate to back. We face a constant barrage of potentially relevant information, evidence, and argument—far more than we could assess in any conclusive manner.... [T]o know that we fully understood something, we would need to conduct an exhaustive and thorough investigation." And that's pretty much impossible, given how complex our world is.

This is likely different than in the past. For our ancient ancestors, the "aha" feeling of clarity was rather reliable because their information quests were straightforward. They either found food or didn't. They had shelter or didn't. They were the high-status leader or they weren't. So we evolved to trust our "aha!" feeling, the feeling of clarity.

But in most cases today, we can't necessarily choose the best information. So we use, Nguyen wrote, the fast and loose "aha!" feeling of clarity to make rough estimates that we've done enough thinking and have made good decisions. But this "aha!" feeling calls off the search before we discover flaws in the information.

For example, the flat-earth conspiracy theorist Mark Sargent told CNN about his views, "You feel like you've got a better handle on life and the universe. It's a lot more manageable." Another flat-earth conspiracy theorist, David Weiss, said, "When you find out the earth is flat ... you become more empowered."

Most of us aren't buying into flat earth. But we probably fall prey to easy information all too often. Or we get cheap, meaningless "aha" moments from silly information online, like whether or not we should take our shoes off at home or how we should hang art. When I asked Nguyen what we can do about it, he compared it to food.

"It's easy to make delicious food if you give up on its nutrition," he explained. "The same happens with truth: it's easy to make seductively clear ideas if you don't care about truth and nuance."

He told me that we should question any information that quickly and easily delivers us a sense of clarity. That "aha!" feeling. We should actually use that feeling as a cue to look for details for how the information might be wrong. Like metaphorically calling the Vatican, we need to be reasonable with this or we'll go mad. But it's critical for topics we want to understand.

Vande Hei told me the same. "I think this is where seeking out ideas that contradict your own comes in," he said. "You don't want to just be reinforcing a preheld belief."

After speaking with Nguyen and Vande Hei, I realized my own "online brain" was acting like a middleman preventing me from having genuinely new experiences. We rarely enter completely unknown worlds anymore. Like, I can't do anything without googling it first.

To take a simple but omnipresent example, a recent survey found that the vast majority of people don't go to a new restaurant before looking at reviews online. Our informavore brain kicks in, and we want to know everything we should expect and how we should order. So we consult an online middleman, like Yelp.

The same goes for watching movies and TV shows, reading books, traveling, buying products, and more. If we can search and scout something beforehand, we're more comfortable taking the leap.

But the problem is that our middleman doesn't always have our best interests in mind. Researchers at the University of Pennsylvania called online reviews a dual-edged sword. In theory, they're great if they can help us make more informed decisions. But the trouble is, they wrote, "there is a systematic problem with many online reviews—they tend to over-represent the most extreme views. . . . This makes it hard to learn about true quality from online reviews."

Let's take, for example, restaurants and how and why people might review them. When we review, we boil down a diverse experience into a star value. But people have different reasons they may like or dislike a restaurant.

For example, someone in a hurry might value expedient service, while someone hoping for a long and meandering experience might be annoyed if their food came out too fast. Someone else may value the restaurant's ambience, while another may just want good-tasting food regardless of if the restaurant has the same aesthetic quality as a pop-up H&R Block tax center. Some might want larger portions, while others might want small plates. So when we write a review, we're often reviewing based on what we most value in a restaurant. But when we read a review, we assume that the reviewer has the same values as we do.

If our person in a hurry got her food exceedingly fast, she'd give the restaurant five stars. Conversely, she'd give it one star if the food was slow to arrive. But if we don't care how long it takes for the food to come out, then her review misguides us.

Perhaps most importantly, our online middleman also alters the rewards we can get from discovery. Searching everything we do before we do it kills new experiences.

Remember how Zentall said that we get deeper rewards from things we had to work harder to achieve. When we look up, for example, restaurant reviews, it kills wonder and dissolves the deep rewards of exploration, unpredictability, and discovery. It would be like if Laura Zerra used maps that told her where valuable antlers and skulls would be rather than playing the game herself. Or if we knew beforehand how the slot machine reels would fall.

When I think of all the restaurants my wife and I love most, none came through the internet. We found our favorite Vietnamese restaurant by chatting with a Vietnamese speaker about where she eats most often and why. We first stumbled on our now favorite Mexican restaurant while driving in an old part of Las Vegas and saw a restaurant sign featuring a cartoon of an old Mexican woman and the word "tamale." So we took a gamble.

This idea, of course, doesn't just apply to restaurants. It applies to all kinds of experiences. Each time we do something truly new and unknown, we can be the hero in our own small, everyday hero's journey. It sounds somewhat silly. But humans have always drawn meaning and deep reward from entering the proverbial wilderness and finding the greener grass. We had to in our world of scarcity.

But today entering the unknown can get us out of a scarcity loop by leaning on the second and third things that can push us out of it. It stops the unpredictable rewards of cheap information and shifts them to something more true and meaningful to us. It also interrupts quick repeatability, shifting fast meaningless information to slower, more memorable information.

It's shifting an online scarcity loop into a real-world, active abundance loop. Just as Zerra searches for antlers without an antler map, we can use the loop to build positive habits and experiences. And, in turn, live life in a more authentic, wondrous way.

This isn't to say that exploration via the internet isn't useful and that you should never use Yelp or Rotten Tomatoes or any other review sites. But we do need to understand its limitations, filters, and algorithms and what those might lead us to miss. Putting in a little more work may create an abundance loop. Some technologists are realizing this and considering it deeply. How can we balance new technology with what's always made humans healthy?

I called John Hanke. He's the computer engineer who led the development of Google Maps and Google Earth.

After uploading the entire globe into Google Earth and Maps, Hanke wanted a new project. Google at the time was entering into augmented reality, known as AR. AR combines computer-generated content with the real world to create an interactive experience. Hanke was the world's foremost mind on mapping software, but he'd gotten his start in tech by designing games. His first money as a tech entrepreneur came when he was a teenager living in Cross Plains, Texas, population 982. Hanke coded an Atari game, which he sent into a magazine that sold reader-designed games.

But as a West Texas kid, Hanke was also involved in sports, scouting, and 4-H. He noticed that the combination of outdoor exercise and exploration and being intimately involved with the natural world enhanced his body, mind, and creativity. The tech stuff was easy for him and he enjoyed it. But he also realized that for optimal health and happiness he had to offset his screen time with active time outdoors.

With that idea in mind, Hanke founded Niantic in 2010. It was a sincere attempt to combine technology with his notions on the benefits of outdoor exploration, activity, and community.

"We built a lot of prototypes of apps that were related to your location on a map and information about the area you were in," Hanke told me. "Our first app was called Field Trip."

Field Trip launched in 2012. To understand it, think of the last road trip you took. You probably passed a handful of highway signs that said, "Historical Site: 1 Mile" or "Cultural Site: 2 Miles." Or perhaps you passed a cool old building. And it piqued your interest. But not enough to stop. So you passed the site and missed out on what was likely some fascinating local knowledge. The Field Trip app was like having the constant companionship of a well-informed tour guide.

Hanke said that when you'd get close to an interesting site, "Field Trip would tell you about it. We wanted to highlight these hidden gems of the world." So you might be walking your dog and the app would alert you that the house you're passing is a Frank Lloyd Wright–designed home and give details on its architecture and history. Or it might explain that the corner you're standing on is an important site in the American Revolution.

Soon after, Hanke realized his son was living a different childhood than he did. "My son at the time was eleven years old and loved video games, but he wasn't going outside much," Hanke told me. "So I was thinking, how can I harness my son's interest in video games, but get him outside exploring and moving more?"

Enter his next game, called *Ingress*. "The idea was, let's take all these data points and interesting places from Field Trip and build a game around it," said Hanke.

Ingress uses smartphones and GPS to layer a game world over the real world. Think of *Ingress* like the game capture the flag but released upon the entire world and with a science-fiction backstory. Players pick a team and "capture" new areas by walking to them.

The game became a cult phenomenon, with ten million downloads. Its players were rabid. Niantic began, Hanke said, "getting messages from people who said they were sedentary before and the game had led them to start taking twenty thousand steps a day."

One of those rabid cult followers was Tsunekazu Ishihara, the head of the Pokémon Company. Pokémon is the highest-grossing media franchise worldwide, with more than $105 billion as of 2021. It's generated more revenue than Mickey Mouse and Marvel and Star Wars. A partnership bloomed.

"So then we rolled over a lot of the same ideas from *Ingress* into a new game," Hanke told me. It used GPS data to place virtual Pokémon in real-world, interesting places. Players walked around using the app to locate, catch, train, and battle Pokémon that appear on-screen as if they were in the player's real-world location. They called the game *Pokémon Go*.

It's popular in the sense McDonald's is popular. Within two months of the game's release in 2016, it had been downloaded more than 500 million times. It's now well crossed the billion-download mark.

"Why do you think it's a hit?" I asked Hanke.

Hanke knows games that leverage the loop draw us in. "These loops mimic real-life reward behaviors," he told me. "So you do something; then you get a reward and you get a rush for getting the reward. In real life, it might be that you get food or do something you need to survive and you're biologically evolved to get a reward from that. Video games kind of copy that loop. Except with normal video games, you get that boost just sitting on the couch: you've leveled up, you feel more powerful, you feel great. And you want to do this more. But it's sort of cheating because you're not actually doing things that are helpful to you.

"By moving that game loop out into the world, we're combining the sort of fake boost you get from a video game with real-life things that are actually good for us," Hanke told me. "So you catch a Pokémon, and you feel good. But you may have also walked a kilometer outside to catch that Pokémon. Or met up with five friends because you wanted to do a raid and it takes multiple people to beat a Pokémon boss. So you're in this highly engaging game loop, but meanwhile you've had face-to-face social time, you've collaborated with people, you've exercised outdoors, you've done these other behaviors we evolved to do to stay healthy and happy."

You're also exploring new parts of the world, Hanke said. "We learned from the Field Trip app that we didn't want to pick random spots to send people for Pokémon. Like, we don't want to point someone to a Walmart parking lot. We wanted to send them to historical markers, or public art installations, or really interesting local businesses. We get a lot of messages from people saying something like 'Hey, there was this historical mill in my town down by the creek and I didn't know about it. And I ran down there to catch a Pokémon. And I found some interesting history about my town and saw and learned about a really cool site,'" Hanke said.

Pokémon Go exposed many couch potatoes to the outdoors and exercise. "Lots of people are just not intrinsically motivated to exercise for the sake of it, and that's totally fine. But these people get left out" by our current exercise complex, said Hanke. "So our games like *Pokémon Go* reach a lot of those people and give them a better reason to get outside and move and be social. When it gets to people and unlocks that, it can be so transformative."

Scientists at the London School of Economics studied how the rise of *Pokémon Go* affected depression rates. They looked at mental health search data before and during *Pokémon Go's* peak use. The scientists wrote, "We argue that the introduction of Pokémon Go, a mobile game that encourages outdoor physical activity, face-to-face socialization, and exposure to nature, may alleviate non-clinical forms of mild depression for users playing the game."

This wasn't news to Hanke. He told me that Niantic frequently hears from people at its community events who have lost more than a hundred pounds, improved their fitness or mental health, or made new friends playing *Pokémon Go*. Its players have collectively walked more than ten billion miles and counting. "It's been the nudge that some people need to do healthy things," Hanke said.

Hanke engineered the video game equivalent of when parents sneak healthy vegetables into their kids' mac and cheese. It's shifting a scarcity loop to an abundance loop—one that helps people.

He recognizes big tech's great power and responsibility using the scarcity loop. "Technology is here. It's not moving away," Hanke told me. "So it's up to us to try to shape it into something that's helpful to us and not destructive to us. If we just leave it alone and let whatever happens happen, lots of bad things are going to occur. It's going to take huge amounts of conscious effort repeated over and over forever to keep it under control."

He said, "This current manifestation of the internet and consumer technology is insidiously powerful because of the way that it can mimic these loops that evolved for other things and were necessary for our survival. They're now essentially being co-opted by a sort of fake system. The tech systems have gotten so good at mimicking these ancient reward pathways. I can tell you that it's absolutely scary when you apply artificial intelligence to that. You can go into an app like TikTok, and with a few clicks it's going to know exactly what's going to trigger that reward release in you to get you to watch more and more and more and it gives you more and more and more of that."

TikTok is currently the best at leveraging the power of this loop. "The end result can be people sitting in a room all day swiping through videos. Not eating well. Not getting an education. Not meeting people. Just sort of existing."

There will be new apps down the road that sharpen the scarcity loop even more than TikTok. But also apps that leverage the loop to lead us to do things that are good for us. Like Strava and Sandlot, which promote outdoor exercise often done in groups. Or iNaturalist, which has us go into nature to discover and learn about different plant and animal species. It's ultimately up to us to be aware of when and why we're falling into the scarcity loop, and look for ways to shift it to an abundance loop.

"We often don't realize to what degree our behaviors are not explicitly conscious choices but are driven by these sorts of subconscious chemical actions that get hijacked," Hanke told me. "It creates something that could potentially be called addiction. If you don't want to call it that, it's highly incentivized repeated behavior."

The greatest journeys in life are never known or comfortable. Their reward is not their destination nor the food or shelter or virtual Pokémon there, but how we encounter the unknown and uncomfortable along the way. They're opportunities to be the hero in our own story. To develop all the skills we want humans to have. Courage, commitment, adaptability, resilience, and more. And to be engaged on a deeper level while improving our physical and mental health.

E. B. White's book *Stuart Little*, about a little boy mouse who gets adopted by a human family, ends inconclusively.

White later commented that the ending "plagued" him. "Not because I think there is anything wrong with it but because children seem to insist on having life neatly packaged," he wrote. "My reason . . . for leaving Stuart in the midst of his quest was to indicate that questing is more important than finding, and a journey is more important than the mere arrival at a destination. This is too large an idea for young children to grasp, but I threw it to them anyway. They'll catch up with it eventually."

Each time we off-load our quest and remove effortful exploration, we quit the journey.

Remember Maslow's hierarchy of needs. For most of time, we explored the world on foot, searching for all the food, stuff, and information we needed to meet the bottom rungs of the hierarchy.

Our world of abundance means many of us no longer need to do that. It's a good problem to have, yet it presents problems nonetheless.

But also an interesting opportunity. Now that our basic needs are taken care of, we can explore the top of Maslow's hierarchy. Those "self-actualization" needs. That inner wilderness of meaning and happiness felt like a good place to explore next.

The pope, by the way, doesn't draw a fucking salary.

Happiness

Brother Brendan had clear instructions. Before arriving at Our Lady of Guadalupe Monastery, I was to do two things.

First, he told me, "you will definitely want to familiarize yourself with the *Rule of Saint Benedict,* which is central to our monastic life." It's a twenty-thousand-word manifesto that guides monks living in a monastery. Chapter 4 of the *Rule* laid out seventy-two "instruments of good works." They're pitched as "tools of the spiritual craft" and are basically regulations within the larger book of edicts and precepts.

Some of these seventy-two rules were clearly doable. Like number 3, "not to commit murder." Easy. Or number 4, "not to commit adultery." Even easier. My wife is the ultimate human and the only women within miles of the monastery were nuns at a convent on the grounds. Or number 45, "to be in dread of hell." Hell, if everything I've ever heard about it is correct, seems dreadful. So, sure, consider hell dreaded.

But some of these guidelines were going to be trickier. Like number 39, "not to be a grumbler." Or 27, "not to swear." Or 1, 41, 42, 49, 58, 60, 62, and 72, which all mention the *G* word. God. Which is a word I've been in an on-and-off relationship with for most of my life.

Second, I was to prepare myself for physical work. "Be sure to bring work clothes and boots," said Brother Brendan. Leave your grumblings and soft hands at home and prepare for labor, my son. "Idleness," the *Rule of Saint Benedict* said, "is the enemy of the soul."

No, this would not be one of those monastic retreats where I paid a bunch of money to sit in the lotus position and do nothing for a week. Hoping to gain enlightenment or something like it when I finally got my fifteen minutes with the guru.

Brother Brendan wouldn't even call my visit a retreat. Benedictine monasteries have been receiving guests since their beginning in the sixth century. Chapter 61 of the *Rule* states, "If a pilgrim . . . coming from a distant region wants to live as a guest of the monastery, let him be received for as long a time as he desires, provided he is content with the customs of the place as he finds them and does not disturb the monastery by superfluous demands, but is simply content with what he finds."

Our Lady of Guadalupe Monastery sits at the edge of New Mexico's Gila National Forest. It's a spiritual anchor in the high desert, pinning down the wilderness as it casts northward as a three-million-acre net of rolling rust-colored mountains, deep canyons, and fir and spruce and piñon trees. The nine-thousand-person town of Silver City is roughly thirty minutes away by mountainous dirt road. The closest major airport is three hours, in El Paso, Texas.

Now, why I was visiting Our Lady of Guadalupe Monastery, I couldn't explain with any coherence or conviction. But I think it had something to do with wanting to explore that last line of chapter 61, to be "simply content." I wanted to get a handle on some research I'd come across suggesting that Benedictine monks had cracked the happiness code in unexpected ways.

Academics have been formally studying happiness for decades. Many of these scientists are so confident in their work that they even give specific advice that they say is "necessary" to follow if we want to be happy. Like, if we don't do what they say, we do not pass go and do not collect happiness.

But I'd become slightly skeptical of this happiness research. After reading it, I could see flaws in the study designs, like how the study was set up, whom it surveyed and how it surveyed them, and how it measured happiness. Much like the problems with nutrition science.

Yet it's popular research nonetheless. Americans are seemingly most entranced by the scarcity loop of happiness. We now spend more money per capita than any other country on wellness products. After all, "the pursuit of happiness" is written into our Declaration of Independence as an "unalienable right." We've basically written the scarcity loop of happiness into our defining document—keep pursuing and pursuing and pursuing.

Yet the data shows that we're far unhappier than other developed nations. Each American generation is unhappier than the one that came before it, according to (rather depressing) new research.

But it's not as if the rest of the world were blissed out. All sorts of metrics suggest we are experiencing, as David Brooks put it, a "rising tide of global sadness." Researchers in the U.K., for example, analyzed the lyrics of 150,000 popular songs released between 1965 and 2015. They wrote, "The usage of [the word] 'love,' for example, practically halved in 50 years." The word "hate" didn't appear in popular songs until 1990 and now finds itself in 20 to 30 hits a year. Similarly, the words "joy" and "happy" fell while "pain" and "sorrow" grew. Another group analyzed millions of news headlines and found that they got significantly more negative. Global unhappiness hit a record high in 2021. One extensive study found "negative feelings—worry, sadness, and anger—have been rising around the world."

But like the Tsimane, these monks seemed to have their own cordoned-off living laboratory. How they were living and finding "simple contentment" was at odds with some of the advice from the "happiness labs" embedded in various universities throughout the country.

I began reading about the life and ideas of Saint Benedict. How he'd grown up in a time when Rome and its people had become a case study in the downsides on scarcity brain run rampant. Yet Benedict managed to find a way out of the scarcity brain cycle that was, he wrote, neither "harsh nor burdensome."

Benedict believed that monks and the public alike should have enough in proportion to their needs, but nothing to excess. And that goes for everything: food, possessions, influence, and the like.

He called this "proportion." It's the recognition that every human has different needs and temperaments. Most religions preach moderation or the middle path. But Benedict understood that "moderation" is different for everyone. Enough for one person might be too much for another might be too little for another. Benedict even taught that self-denial and going with too little often stirs up pride, a snooty "holier than thou" attitude. Having too much or too little, he believed, was a distraction from the ultimate goal.

Benedict realized that we're all seeking happiness. That's the capital *G* Goal underlying all our actions. It drives scarcity brain. It drives the scarcity loop. But our common tragedy isn't that we can't find happiness. It's that we look for it in all the wrong places. We look for it—as Benedict noticed in Rome—in material possessions, power and respect, or fleeting pleasures like food and drink. We fall into a scarcity loop believing that *this time* the slot machine symbols will align and we'll score a permanent win.

It was black on the edge of the Gila wilderness. I'd been driving a serpentine dirt mountain road edged by dark pines for nearly half an hour and was beginning to suspect I was lost.

Then my high beams caught two twenty-foot stucco pillars flanking the road. A cross topped each. The pillars swooped out into big curving walls. The pillar wall to the right held decorative tiles showing an image of the Virgin of Guadalupe and square blue letters spelling out:

MONASTERIO
DE NUESTRA SENORA
SANTA MARIA DE GUADALUPE

I passed through to the other side. Out ahead was a building constructed in the shape of an H. Each wing was at least thirty feet tall. One wing's slender, eight-foot-tall arched windows shone with light. Brother Brendan had texted me hours earlier and told me to go to the building with the lights still on.

A monk named Brother Lawrence was inside waiting for me.

He was all teeth and gums, smiling widely and reverently as we shook hands. He was wearing a heavy tan hooded robe that draped to the ground. It covered the type of overbuilt brown leather work boots you might wear to plow a field or invade a small country.

His glasses were round and unframed. Head shaved. If it weren't for the robes, I might assume he was a tech worker who dabbled in power yoga. We jumped into my car and drove to the cloisters a quarter mile down the dirt road.

This complex was covered in light tan stucco and roofed with terra-cotta tile. Rustic Spanish Southwest style. A fifty-foot-tall bell tower rose from a corner.

Brother Lawrence and I walked up a set of small steps into an arched, open-air entryway. He pointed to carved, medieval-looking double doors on the right. "That's the chapel," he said in a whisper. "Matins is at 3:25 a.m. sharp."

"Um, I'm sorry, but what's Matins?" I asked. Brother Lawrence gave me a quizzical look. Most pilgrims who visit the monastery are Catholic in the sense that the pope is Catholic. The only Catholic service I'd ever attended, on the other hand, was an Ash Wednesday session a babysitter dragged me to when I was seven years old.

"Matins is our first worship service of the day," he said.

He then swung open a larger carved brown set of double doors, and we walked into open-air cloisters. Spanish tile was smooth underfoot. The wide, roofed walkways outlined a garden area filled with wild red and yellow roses and an ornate fountain.

We approached another set of medieval-looking wooden doors. Brother Lawrence opened them and led me through.

The room was a long rectangle with twenty-foot ceilings. Hardwood tables lined with hand-carved wooden chairs ran the length of it. Each seat had in front of it a large mug, upon which rested a dish. A six-foot-tall crucifix hung on the back wall, overlooking everything.

"This is where we take our meals," said Brother Lawrence. "Breakfast is at 7:00 a.m., after Lauds, our second worship service, which is at 6:00 a.m. Here's your seat," he said. I even had my own nameplate. "Michael Easter" scrawled across a piece of wood in Old English. I was at the end of one of the tables, seated next to a nameplate that said, "Br. Paul."

Brother Lawrence paused as we were leaving the dining hall. "Oh, and by the way, breakfast is a buffet. So eat what you want," he said. "And we take breakfast standing, so don't sit." He opened the door a little more, then paused again. "Oh, and we eat in silence."

As we walked back toward the chapel, Brother Lawrence quietly pointed out more doors and passages in the complex. There was a kitchen, a couple of meeting rooms, and a massive library. Living quarters on the second floor. Every one of the forty monks who live at Our Lady of Guadalupe Monastery has his own ten-by-ten-foot room. Each contains a desk, chair, bookshelf, twin bed, and closet that holds the monk's four outfits: two black robes for prayer, two tan ones for labor. And the killer work boots.

Brother Lawrence hopped into a 90's-era Toyota T100 pickup, and I tailed him half a mile up the dirt road to the guest quarters. Picture a big, don't-spend-too-much-because-they'll-probably-break-it Airbnb rental property, with a bunch of dorm-style bedrooms and communal every-thing. Walls mostly bare except for a few framed paintings of Saint Joan of Arc or the Virgin of Guadalupe.

"The abbot is a big fan of Saint Joan," said Brother Lawrence as he walked me down a long dark hallway. He opened the door to my room. It contained a twin bed, a wooden chair, a wooden dresser. No TV. No internet. And I had no bars of cellular service.

The plan was to spend a week at the monastery. Embed myself with the monks who pray, live, and work on the grounds, mostly in silence and removed from society.

"See you in the morning," Brother Lawrence whispered. I could hear his heavy boots clunking down the hallway. They crunched the dirt as he crossed the parking area in front of the guesthouse. An engine turned. Tires rolled on gravel and faded into the night.

Thomas Aquinas, the thirteenth-century Dominican monk and philoso-pher, put it this way: "Beneath the multitudinous and even conflicting desires of [people] we can see the one desire which gives unity and mean-ing, force and decision to all human desires. All [people] seek what they seek for one reason: they think it will satisfy them, they believe that the accomplishment of their desires will make them happy.

"Happiness," he wrote, "is the goal of all human activity. The search for happiness is the common ground on which all human desires, all human ambitions meet."

When a child looks longingly through a pet shop window at a puppy, she's seeking the dog because she's seeking happiness. When a miner toils in a coal mine, even though the work is harsh drudgery, he's seeking coal because he believes his pay will lead him to happiness. When a sales executive strives to make the next big deal, she's seeking a commission because she believes the commission will ultimately bring her happiness. When we take that second serving of food, troll someone online, click buy on Amazon again, or do anything at all, we're taking that action because we think it will make us happy. When we fall into a scarcity loop, it's for happiness. Even our worst ideas are a search for happiness.

But why? Aquinas explained that people "have natural desires for whatever is required by nature itself for our well-being."

Scientists starting proving him correct as far back as 1872. That's when Charles Darwin began publishing ideas about the role of emotions in evolution. As humans evolved in rough-and-tumble landscapes, the desires we've investigated throughout this book helped us survive. Each time we'd use a drug or get more food, possessions, prestige, or information, we'd be rewarded with good feelings of pleasure or joy. That, in turn, produced the mysterious and rewarding feeling we call happiness.

Yet the feelings were fleeting, because tomorrow we'd again battle for scarce resources. It was something of a scarcity loop. We'd take an opportunity to improve our lives, feel suspense as we waited to learn the unpredictable outcome, and then experience happiness if we were successful. But this happiness was just a brief hit. We'd again be shot right back into dissatisfaction and longing. Into the anxiousness and suspense of seeking the next fix of whatever it was we thought could bring us pleasure and joy. Repeat this cycle for life.

But this cycle kept us alive. As we evolved, sustained happiness would have killed us. It would have been more of a bug than a feature in our mental hardware. If anything permanently satisfied us, we'd have given up on tasks that were necessary for our survival and died off. And so we lived in a scarcity loop of happiness.

This is why neuroscientists in the U.K. say there's no biological basis for sustained happiness. Scientists still can't point to a single area in the brain that controls happiness. Yet happiness isn't just murky in brain scans. It's also murky in general.

Thinkers going back thousands of years and modern scientists have tried to pin down happiness. To explain what it is and how we can make it last. But we haven't been all that successful.

The dictionary now defines "happiness" as "feeling or showing pleasure or contentment." But if we look to the definitions of "pleasure" and "contentment," they point right back to the word "happiness."

Others have developed their own definitions of happiness. But they're unsatisfactory. The Greek philosopher Seneca said happiness is "[enjoying] the present, without anxious dependence upon the future." Dale Carnegie said happiness is something "governed by our mental attitude." For Emily Dickinson, it was "The mere sense of living." For John Lennon, happiness was a warm gun. It's not surprising that many of these definitions are so different.

Researchers in Singapore discovered that our definition of happiness might depend on where we grew up. The feelings people most cite as "happy" and the situations they say make them happy differ between cultures. The scientists found that in the West the themes that most appear when people talk about happiness are "peppy emotions like excitement and cheerfulness" and self-esteem. In the East, people reference "calmer states like peace and serenity."

And how you experience happiness is constantly shifting across your life based on all sorts of factors seen and unseen. Where you were born, who raised you, what you do for work, whom you hang out with, and so much more. Every moment, thought, and action affects it.

So perhaps it's best that we can't give happiness a firm definition. Considering the elusive nature of happiness, the scientists wrote, "a state of contentment is discouraged by nature because it would lower our guard against possible threats to our survival."

The catch is that today we don't have as many threats to our survival. Our modern world has pushed abundance onto us fast and hard and taken care of our basic survival needs. Yet our brain still continually tosses out our happy feelings as if they were rotting garbage.

The economist Brad DeLong explains that before the Industrial Revolution the comfortable people of the world had to "attain such comforts . . . by taking from others, rather than by finding ways to make more for everyone." In other words, if they wanted to be happy, they probably had to make another unhappy.

But today, DeLong writes, "less than 9 percent of humanity lives at or below the roughly $2-a-day living standard we think of as 'extreme poverty,' down from approximately 70 percent in 1870." That figure is adjusted for inflation. "And even among that [poorest] 9 percent, many have access to public health and mobile phone communication technologies of vast worth and power."

We've turned incredible experiences into everyday features of living. DeLong writes, "So many of us have grown so accustomed to our daily level of felicity that we utterly overlook something astounding. We today—even the richest of us—rarely see ourselves as so extraordinarily lucky and fortunate and happy, even though, for the first time in human history, there is more than enough."

We produce enough food, shelter, clothing, and stuff that no one *has* to be hungry, wet, cold, or without necessities (and when people are without those necessities, it's usually a problem of distribution and politics). Yet scarcity brain, as those U.K. neurologists found, still craves and clings to emotional swings. It can't sustain happy feelings.

The same cycle that helped us survive in the past—happiness followed by dissatisfaction repeated for life—now blinds us to how astonishing modern life is and leads us to chase happiness in all the wrong places.

The research consistently shows that money, power, prestige, food, drink, stuff, status—what Aquinas would call "worldly" pleasures, and what you and I might call the American dream—don't typically lead to lasting happiness. For example, Americans didn't get any happier from 1975 to 1999, even though they became 43 percent wealthier based on per capita GDP. Some evidence suggests that as we've had more opportunities for more of those things, we've become *less* happy. The United States experienced a serious happiness downturn starting around 2015. And the number of Americans reporting that they were "very happy" recently dipped to an all-time low in 2018, before the pandemic.

Perhaps our influx of more of everything we evolved to crave paired with our industrial happiness complex—obsessing about and pushing happiness—is making us feel as if any day that isn't pure bliss were problematic and leading us to chase the wrong things. The U.K. neuroscientists write, "Pretending that any degree of [dissatisfaction] is abnormal or pathological will only foster feelings of inadequacy and frustration." Given our wiring, they write, "dissatisfaction is not a personal failure. Far from it. [It is] what makes you human."

Funnily enough, Benedict seemingly didn't care about happiness. In all twenty thousand words of the *Rule,* the word "happiness" or "happy" doesn't appear at all. But maybe the result of living by the *Rule* produces happiness. At least that's the belief of Alex Bishop, a professor at Oklahoma State University who has studied Benedictine monks and nuns extensively. He studies humans across our life span: what makes us live long, well, and happy.

I spoke to Bishop before arriving at Our Lady of Guadalupe.

"The Benedictines have high life satisfaction scores," he said. "They have higher sense of purpose and meaning. They're higher than the general public. And you see this in other studies on them too. They're pleasantly, well, happy."

And the reasons for this don't exactly jibe with what we've been told about happiness. "There's a lot of confusion and paradoxes around happiness," he told me. "It's so variable. It can change on a dime." Not just our own happiness, but also what we think we know about happiness. "And I think that's where these Benedictine communities become interesting," said Bishop. "It's a rather austere lifestyle."

I woke at 5:30 a.m. I was two hours and five minutes late for Matins, the "night office" of prayer in the chapel. I'd been at the monastery for eight hours and had already broken rules 37 and 38, which tell us to be "not drowsy" and "not lazy."

Black engulfed the wilderness, and a cold breeze was pinging a set of steel chimes hanging on the back porch as I left the guesthouse. A small flashlight lit my way down the half-mile forest-flanked dirt road leading to the chapel. My light outlined piñon pines and juniper trees rising intermittently from the sloping sides of the road. The breeze was pushing a smell of sage. The sky above was a net of white stars, with seemingly no beginning to the sky or end to the trees and hills.

I sat at one of the four hardwood pews. At 6:00 a.m., the monks filed in. They were in black hooded robes and silent. Each knelt toward the altar, then took a position in elevated wooden stalls that ran along the walls beyond the pews.

Then it began.

"Dóminus regnávit, irascántur pópuli: qui sedet super Chérubim, moveátur terra," they chanted, the music echoing off the stone walls and hardwood roof. "Dóminus in Sion magnus: et excélsus super omnes pópulos."

This monastery is one of the few remaining that is 100 percent obedient to the *Rule*. Chapter 16 asks that monks pray and chant together eight times a day every day without exception.

Matins at 3:25 a.m. Lauds at 6:00 a.m. Prime at 7:45. Terce at 9:30. Sext at noon. None at 2:00 p.m. Vespers at 5:00. Finally, Compline at

7:00. All these sessions chant from the Liturgy of the Hours, a series of prayers written to be sung to praise and thank God.

There are about twenty thousand more Benedictine monks like them, spread throughout four hundred monasteries worldwide. They've all retreated to give their lives over to something larger than themselves.

In the year 494, fourteen-year-old Benedict moved to Rome to study. The city and Roman Empire were in a free fall. Historians say decadence was a key driver of the downfall. According to Pope Pius XII, Benedict "noticed heresies of all manner ... private and public morality were crumbling and very many, especially the fine elegant youth, were sadly sunk in the mire of pleasure." Benedict wrote that Roman society was "dying and it laughs."

This is a phenomenon noticed by the fourteenth-century Islamic scholar Ibn Khaldun. He found that societies fall in a similar pattern. The founders of a society work hard under harsh conditions to establish it. These harsh conditions require and develop social cohesion, as we might see today in military units. But, over time, generations follow that become too far removed from the original hardship and work ethic it took to attain prosperity and maintain a society. These generations become soft and comfortable and adopt an attitude of entitlement. This ruins the society from within while making it prone to attack. Which is exactly what happened to Rome, albeit over about five hundred years.

Benedict put it this way: just as pride comes before the fall, so does excess.

So Benedict bailed. Pope Pius XII wrote that Benedict "willingly bid farewell to the comforts of life and the charms of a corrupt age, as well as to the enticing and honorable offices of a promising future ...; leaving Rome behind, he sought out wild and solitary places where he could devote himself to ... contemplation."

"Contemplation" is also a slippery word.

The word appears across spiritual and intellectual disciplines. Plato saw it as going above knowledge to understand what is and always will be *good*. In Psalm 27:4, it's "behold[ing] the beauty of the Lord." The Jewish philosopher Maimonides defined it as a spiritual act of recognizing moral perfection. The founder of the Baha'i faith, Bahá'u'lláh, thought it was reflecting on beauty, God, science, and arts. The Islamic prophet Muhammad viewed it as considering life, its meaning, and Allah's and humanity's welfare.

So the word has deep spiritual connotations. Secular definitions all suggest it means trying to wrap our head around what is larger than ourselves.

So how do we actually "contemplate"? Benedict did it by escaping to a cave up in the hills of Subiaco, Italy. He was following the footsteps of early Christian hermits like Paul of Thebes and Anthony the Great, who fled into the Egyptian desert in the year 250. These Desert Fathers and Mothers believed that solitude, austerity, and sacrifice were the highest form of finding a higher power.

In the cave in Subiaco, wrote Pius XII, Benedict "strove for three years with great fruit to acquire the perfection and holiness . . . He made the practice of shunning all earthly things to seek alone and ardently heavenly things; . . . In this way of life he found such sweetness of soul that all the former delights he had experienced from his wealth and ease now appeared distasteful."

Word spread about this long-haired, long-bearded mystic who rejected the comfortable world, suffered the solitary hardships of the wilderness, and came out on the other side a sage. Wanting nothing. Needing nothing. Just cool, clearheaded, and content.

Romans began traveling to him for advice. These people all felt as if something were missing from their lives. All their striving to complete the ancient Roman checklist of getting ahead—get rich, buy stuff, make a name, wield power, fulfill the next impulse—hadn't necessarily gotten them to where they thought it would.

Benedict had good news for these people. From his three-year austere experience alone in the cave, he'd learned that solitude is enlightening but we probably don't have to go to the extremes he did. Sort of like, "Well, I'm glad I did that so I learned you don't have to." To help others, Benedict began founding monasteries.

He'd establish a monastery, recruit twelve monks, make sure everything was in good working order, then go build another. And so on until he ended up with twelve total monasteries. They served as places where monks could live the monastic life and help the public do a bit of their own contemplation.

Benedict wrote the *Rule* in the year 516 to guide those monasteries. The *Rule* eventually became the standard guidebook for Western monastic life, whether monks were Benedictine or not. That's because, in true Benedict style, the *Rule* was "neither harsh nor burdensome.... [H]e tried to govern his disciples by love rather than dominate them by fear," wrote Pope Pius XII.

Pius XII explained, "The community life of a Benedictine house tempered and softened the severities of the solitary life, not suitable for all and even dangerous at times for some." The *Rule* has been called genius in its adaptability, prudence, discretion, and balance of severity and mildness.

Benedict inherently understood that cut-and-dried rules often leave too many people high and dry. The monks have a saying, "To some, more is given. To others, less." For example, a monk who thrives off social interaction may be assigned to a job that allows him to talk more. Days of fasting are also individual; the idea is to take whatever amount of food feels like a sacrifice for you. And most of the monks avoid meat, but some brothers, especially the young who work the physically hardest jobs, eat it to fuel their labor.

This regimen of finding and doing with enough, Benedict taught, allows us to focus on what truly matters: time where we discover that something larger than ourselves isn't absent from everyday life. Benedict believed we find higher purpose and satisfaction in helping others, experiencing and making creative works, learning new things, balancing time in solitude and with others, and awareness in nature. And, most important, contemplating whatever big eternal mystery we think is the Big Eternal Mystery and letting that guide us.

Benedict's philosophy on life can be summed up by the phrase "ora et labora." That's Latin for "pray and work." It's the motto of Benedictines.

The monks practiced the faith, helped others, and labored together at a unique trade. For example, one monastery was filled with expert dairy farmers. Another with metalworkers. One with shepherds who could weave a mean wool cloak. Then the monasteries would all trade what they'd created so they could sustain each other and also sell to the public to fund themselves, no handouts needed.

The sun was beginning to rise over the Gila, and light was entering the chapel. It was catching circular patches on the wooden platform where the monks stood, areas polished smooth by years of the men kneeling in prayer.

The monks had been chanting for thirty minutes. "Numquid non ego Dóminus, et non est ultra Deus absque me?" Their voices were mixing with the early morning chorus of crickets and the crowing of a rooster. Occasionally, one or two monks would solo a verse, but most times the monks were like the Eagles, all harmonizing together. And they were rarely stationary. The verses called them to lean back into their hardwood stalls, step out to stand tall, bow, or hit their knees.

I'd expected a bunch of men who would qualify for AARP. I'd read statistics about fewer young people entering organized religion. But most of the monks were younger, some as young as twenty. Brother Lawrence told me the average age at the monastery is thirty.

An older monk chanted a final prayer. Then the group filed out of a side door. I left by the main door, and one monk—tall, bald, and robed— opened the double doors leading to the cloisters. Brother Lawrence motioned me to the dining hall door.

I entered the hall. The smell of freshly brewed coffee filled the air, the same roasty scent of a classic roadside diner early in the morning. Coffee is this monastery's "labor" that pays the bills. They roast twelve hundred pounds of beans a week but have ordered a new roaster to increase their production.

A handful of monks stood silently in front of their chairs, eating from mugs. I joined the few shuffling toward the long hardwood table that held the breakfast spread. Most food here is picked, produced, and prepared on the grounds. There was bread, yogurt, milk, granola, fruits, eggs, and honey.

The monks took meager portions. Rules 14 and 36, after all, tell us to "love fasting" and "to not be a great eater." So I ladled a bit of yogurt made on-site into my mug and sprinkled some granola atop it.

We all stood eating in silence. I sipped the coffee. I am, admittedly, a bit of a coffee snob (and I truly hope coffee is the only thing I'm snobby about). I get into roasts, the origin of beans, tasting notes, and the like.

This coffee was none of the hipster, third-wave, snobby stuff I was used to. It was *coffee*. Just good coffee. No notes of orange zest or deep caramel sweetness or anything like that. It was effortless, subdued, and solid. Much like the monks themselves.

After a few more sessions in the chapel and lunch, I was sitting on the front porch of the guesthouse when a 1989 beat-to-hell Chevy Silverado 2500 sped up. One of the brothers hopped out and emerged through the dust he'd created skidding the pickup to a stop.

He was around six feet three and 190 solid pounds. If he wasn't a monk, I could picture him on a construction crew. He seemed to be just under thirty years old. Maybe twenty-eight.

He was wearing tan work robes and Ray-Ban Aviators. Hair cut in what's called the tonsure style, with his head shaved everywhere, except for a ring that runs the circumference of his head. It's an arresting look I hadn't seen outside old monastic paintings and lithographs. Most of the monks here have shaved heads, but those who are on the path to becoming priests and teachers still rock the tradition of the tonsure cut.

He stuck out his hand. "Hi," he said, "I'm Brother Cajetan." Monks receive a new name when formated as monks. The abbot, the monastery's head honcho, names them after a saint. Brother Cajetan is named after Saint Cajetan, the patron saint of the unemployed, gamblers, and good fortune. The gamblers in my hometown of Las Vegas often carry coins with Saint Cajetan's likeness.

"I hear you'll be working with me today," said Brother Cajetan.

"Sounds good to me. What are we working on?"

He pointed to a big pile of rocks near the guesthouse. "We're going to pick up that pile of rocks, and we're going to move them there," he said, pointing to a spot twenty feet away.

It took me far too long to realize that he was joking. I chuckled. "Well, I guess that would be like a workout," I said. "You know, just doing something physically hard for the sake of it."

Brother Cajetan looked at me pensively. "Yeah, no," he said. "That would be silly and useless. We're going to do something more productive than that."

Then I was sitting shotgun in the old Silverado, which was careening down the dirt road toward the cloisters. Brother Cajetan was driving *Dukes of Hazzard*–style. All gas, no brakes, pushing the Chevy's engine and stiff shocks to the brink over rocks and ruts and potholes. A big cloud of dust exploded and expanded behind us. And I could nearly hear that opening E chord and Waylon Jennings singing the theme from *The Dukes of Hazzard*, "Just the good ol' boys / Never meanin' no harm / Beats all you never saw / Been in trouble with the law since the day they was born."

Then Cajetan and I were standing in a twelve-by-twelve-foot garden attached to the side of the monastery. Our project did happen to be moving rocks, but with more purpose.

"This garden is unproductive, so the abbot wants to get rid of it entirely," said Brother Cajetan. "We're going to shovel all the dirt and rocks into buckets and move it all up to the rose garden in the cloisters."

As we started shoveling, the sky darkened and dampened. Rain poured. I began a mental prayer for this work to be over soon. *Ora et labora*, indeed.

Benedict devotes an entire chapter of the *Rule* to labor. He writes, "they are truly monks when they live by the labor of their hands," and so "the brethren should be occupied at certain times in manual labor." Some days these monks work two hours and other days four or more. Benedict was wise on the nature of work.

Four hours of focused work a day seems to be the sweet spot for productivity. Our greatest thinkers throughout history like Charles Darwin, Charles Dickens, Ingmar Bergman, and G. H. Hardy swear by four hours. "Four hours creative work a day is about the limit," said Hardy, one of history's greatest mathematicians.

Science has proved Benedict and Hardy and all those others right. In the 1950s, a group of scientists surveyed a range of the country's researchers about their work habits. They found, quite paradoxically, that more work didn't lead to more productivity. The researchers who worked hard twenty hours a week pumped out the most scientific articles. They released double the number of studies than did their counterparts who spent thirty-five hours a week in the lab. But those thirty-five-hour researchers were far better off than the ones who spent sixty hours in the lab. The latter group produced the fewest articles.

Follow-up research suggests that working four hours daily allows us to find the sweet spot between hard work and adequate rest. We can work intensely for four hours and get a lot of good work done. Less than that and we leave performance on the table. More than that and we become likely to overdo it. To get injured by physical work, to get burned out by mental work. And this cuts into our future workdays.

The nice part is, we can sustain four hours of daily work seven days a week. And despite that whole rule about Sunday being a rest day, many monks work on Sunday. Or else no one would eat and the farm would go to hell.

Accommodating guests and sustaining themselves are just the practical parts of this work. As we shovel, Brother Cajetan is giving me a sort of sermon on the mount of dirt about the higher calling of labor. Genesis 1 and 2, the opening chapters of the Old Testament, tell us that humans are made for labor. And the point of work isn't to finish it. It's to do it. Work, he said, can be a form of prayer and devotion. It's a means to get closer to something larger than the job at hand.

Any work done at the monastery goes to a higher good. It allows the brothers to get closer to God, to help the surrounding community, and more. One Benedictine described labor as a "call to awareness and mindfulness." They wrote,

> Work can express our relationship with the world around us. As such it will not always be rewarding, just as life isn't. But if I believe [work] can be no more than instrumental (paying bills), I will never notice what else it could be. If I simplistically equate self-satisfaction with work's value, I'll miss what else work, even tedious work, can yield. Conversely, looking for deeper dimensions in work may motivate me not to be exploited on the job, not to over-work, not to reduce my life to how much I can accomplish or equate my [self-worth] with my work-contributions, earning power or talents. After all, I own none of those; they are visitors and they may not always call. A day will come when they visit me no more.

Please remember that this Benedictine isn't expounding on, say, performing miracles. Monastic work is rarely sexy. It's lots of manual labor and chores.

"So this job we're doing here," said Cajetan, projecting his voice through the rain, "it's a metaphor. It's a metaphor for persistence, for patience, for being willing to suffer for the common good. To love and get closer to something larger."

It's often argued that many modern jobs are soul suckers. For example, in 2013, the famed and late anthropologist David Graeber proposed a theory he called "Bullshit Jobs." He described it in a 2018 best-selling book of the same title. Graeber claimed that anywhere from 30 to 60 percent of all jobs today are "bullshit" and that the ratio of bullshit jobs is increasing over time.

He wrote that these jobs create "profound psychological violence." But when a team of scientists from the University of Cambridge analyzed all the data, they found that, yes, if you believe your job is useless, you will likely experience poor well-being. Yet they wrote, "The proportion of employees describing their jobs as useless is low and declining and bears little relationship to Graeber's predictions." In fact, many of the jobs Graeber identified as being "bullshit jobs" reported the highest levels of job satisfaction.

See, that's the thing. No job is useless so long as you're treated well and can realize that it's probably helping someone somewhere.

"Benedictines are able to gather meaning from their work," said Bishop, the Benedictine happiness researcher. He told me this reliance on work into old age—drawing meaning from simple tasks—may explain why monks tend to age better, as measured by how healthy and satisfied they are.

I enjoyed the opportunity to talk to Brother Cajetan. Nearly all my time until working with him had been silent. And that's the norm here. Silence is sanctified.

Rules 53 through 55 have to do with noise. Mouth noises. Talking. For example, rule 53 is "not to love much talking," while 54 is "to not speak useless words." And from 8:00 p.m. to 8:00 a.m., the monks enter what's called the Grand Silence. As one nun explained it, "All speech, except in cases of grave necessity, is forbidden at this time."

One French monk in the early twentieth century gave a case for silence like this: "Our silence is not just emptiness and death. On the contrary, it should draw ever nearer, and bring us nearer, to the fullness of life." Humans have increased the world's loudness by roughly fourfold, and research does back the benefits of time in silence. It's been shown to decrease stress and improve focus and productivity.

It's okay to talk during work. And Cajetan and I talked. But it wasn't small talk; each topic was rich with meaning. "I joined the monastery because there is no higher calling," said Brother Cajetan. "And if it doesn't work out, well, it's not like I wasted my time if I spent it giving my life to God."

The rain had shifted to a sprinkle. At one point, I kid you not, Brother Cajetan stopped shoveling and looked at me and said, "We do this out of love." Paused, thought, and finished. "And to love is to be vulnerable, and to be willing to be wounded." Then he just kept on shoveling as I stood like an idiot trying to peel back the layers of insight in that line.

Toward the end of our two hours of labor, Cajetan and I brushed ourselves off. He led me up through the cloisters. The smell of roasting coffee became stronger with each step. He opened two double doors. A big black coffee roaster was running at full throttle. Brother Lawrence stood by it, writing a few notes on a clipboard.

Our Lady of Guadalupe Monastery first funded itself by creating furniture. "It was beautiful, old-fashioned Southwest-style furniture," said Brother Lawrence. "It was hand carved and made of mahogany or mesquite. Everyone loved the work. The world loved it so much they began demanding more of it. Then people told us we had to have the pieces finished by arbitrary deadlines, such as when a person was finishing a house or wanted to pick it up."

He paused and smiled. "But we have a different schedule than the modern world and a different calling. There is a reason *ora* comes first in 'ora et labora,'" he said. "So we gave it up and now do coffee."

There is, apparently, an expression in French to describe the work style here. It is "un travail de bénédictin." It means "a Benedictine labor." It describes, as the academic and essayist Jonathan Malesic put it, "the sort of project someone can only accomplish over a long time through patient, modest, steady effort. It's the kind of thing that can't be rushed. . . . It's work that doesn't look good in a quarterly earnings report. It doesn't maximize billable hours. It doesn't get overtime pay."

Anything a Benedictine monk produces is produced *well*. Built to last. Form and function meet. We see this in the abbeys Saint Benedict founded fifteen hundred years ago that are still standing today and the ornate woodworking that adorns the chapel at Our Lady of Guadalupe Monastery.

Brother Lawrence explained the appeal of working with our mind and body when he said, "It's an active rather than passive use of time and attention. Falling into the internet or TV is passive." A manual hobby can create an abundance loop because it's active and rewarding in a way that produces something tangible.

And so it is with producing coffee, said Brother Lawrence. It allows the monks to take the same methodical, hands-on approach to production that they did with furniture. But they can make more product faster. "We can make coffee without it really interrupting our prayer," said Brother Lawrence. The coffee operation started when one of the brothers traveled to Brazil to help establish a monastery. Along the way, he met a man who owned a coffee plantation that was producing award-winning beans that were then only available in Brazil.

The coffee and various other projects cover the bills. But the brothers don't own anything. Anything at all.

"We take a vow of poverty," Brother Lawrence told me. "So I personally don't own anything. I don't have a bank account. I have nothing to my name." His everyday items are basically out on loan from the monastery. Laura Zerra would be proud.

Benedictine monasteries are, in fact, a radical attempt at communal living. But the fact that Benedictines have accomplished what the hippie communes of the 1960s couldn't isn't advertised. One Benedictine scholar explained that a monastery "commands personal poverty, but does not elevate being impoverished as if it were some sort of virtue simply to be poor."

Going without for the sake of it, as one English bishop in the early twentieth century explained, is "a weakness, not a strength. The only purpose, the only justification [for going without] . . . is that [a monk] may be more free." By detaching from material things, the monks are freer to attach to a bigger thing. If everything is everyone's, nothing takes on any special significance beyond another tool for the job. It's gear, not stuff.

The chapel bells rang, calling us. Brother Cajetan told me our job tomorrow will be more digging. As the book of Exodus says, "Let more work be laid upon the men, that they may labor therein."

My first day was followed by more days of the same: wake early, go to the chapel, eat a meal, break, go to the chapel, eat a meal, break, do work, break, go to the chapel, eat a meal, go to the chapel, sleep. I found something like serenity in the repetition. I felt calmer and more connected. Not socially, but to myself. Living that schedule also gave me even more respect for the monks here. Especially the twenty-year-olds. They all plan to die here.

Imagine someone asking you what you're doing on any random calendar day decades from now and being able to provide that person with an accurate schedule down to the minute.

Their friends and family members will go to college, get married, buy houses, have kids, suffer the highs and lows of employment, parenthood, sickness and health, and grow old. Meanwhile, the monks here will pray and work. As it is now and ever shall be.

During breaks, I'd sit at the guesthouse and read or stand in the one ten-by-ten-foot patch of earth in the chapel parking lot that got cell service and send my wife a text.

I was enjoying learning more about Catholicism. The faith is the foundation of Western culture and its morals and myths, and I'd never cracked a Bible.

I grew up in a town that was at least 95 percent Mormon. I was a five percenter, even though half of my extended family was Mormon.

Being a nonbeliever in the land of the Latter-Day Saints came with highs and lows. Most community functions ran through the local church wards, so I spent a lot of time alone. Once, a girl in junior high publicly dumped me when she discovered I wasn't Mormon. But situations like that played out in quieter ways often. And I also saw the faith exclude people for reasons they couldn't change, like homosexuality. That seemed strange to me as a kid. (The church has since become more accepting of LGBTQIA+ groups and gay marriage, although most people in the LGBTQIA+ community say the church has more work to do.)

But it wasn't all exclusion. Some in the community involved and helped me. For example, when I was fifteen, a Mormon scout leader noticed me becoming less active in scouts. Scouting ran through Mormonism, and I at the time was becoming more interested in girls and electric guitars. This guy seemingly made it a personal mission that I earn my Eagle Scout ranking. And I did.

I could also see that my grandparents' Mormon faith undoubtedly made their lives richer. It gave my grandfather purpose after he retired in his fifties. It introduced them to friends and gave their network a common worldview. It gave them a good excuse to quit drinking and smoking.

Probably for this reason, my mom, a single parent, thought some exposure to spirituality would be good for me. She'd drag me to a Unitarian Universalist church when I was a kid. In that faith, the only spiritual rule seemed to be that there were no rules apart from the obvious. Don't murder, steal, or be adulterous. Got it. Beyond that, people were instructed to accept everyone and believe what they wanted. In Sunday school, kids would learn about a different religion each year. I dug the acceptance thing. But the experience also felt something akin to a Grateful Dead concert meets a Theology 101 course. The spirituality stuff didn't entirely stick.

By my late twenties, at the height of my drinking, a friend asked me about my thoughts on religion and God. I told him, and I quote, "If someone needs some set of hocus-pocus stories and rules and the promise of a good afterlife to be a decent human, that person sucks." I said that as I was on my way to a bar to black out and burn down my life.

Imagine having your head that far up your ass. In her work *Glittering Images,* the intellectual and cultural critic Camille Paglia wrote, "Sneering at religion is juvenile, symptomatic of a stunted imagination."

Paglia's view now seems about right to me. I uttered my ignorant take on God and religion the night before I got sober. The next day I awoke surrounded in chaos. This happened often. But this time something wild happened. I could see that my drinking was going to kill me early. It was as if a portal opened for a moment. I realized that if I were to walk through it right then, despite knowing that it would come with all sorts of physical and mental suffering, I might have a chance. That was about a decade ago and today I'm sober and, as of this writing, not dead yet.

I've come to believe it's entirely possible that there was something larger at play in that moment. It was almost as if God heard my bullshit the evening before and said, "Don't believe? Watch this, sucker."

I now think there's something much larger than myself out there. I call it God, because language is limited and that word is convenient. I've also come to realize I don't have to fully understand my idea of God. As I've alluded to, I'd also be lying if I didn't admit that my belief comes in waves. Which is why I was somewhat hesitant when the *Rule* mentioned God 122 times. I'd have to confront that shaky foundation and talk about it with men who are bedrock solid in their belief. I actually felt slightly jealous that the monks could believe so wholeheartedly.

As a science journalist and professor, I believe science offers us many answers. But through all my time on the ground speaking to people across the spectrum of living, I've learned that some of the most important things about being a human—the things that most change a person's and a community's life for the better—can't be measured. For thousands of years, we got to these "things," these ideas, through myth and ritual. Then science came in and started to help explain why certain ideas and traditions work or don't. We started measuring in numbers and data and figures. Somewhere along the way we devalued those most important unmeasurables.

In my work and thinking, I now think of it as balancing science with soul. Too much science, and we lose the most important aspects of the human experience. Too much soul, and we can lead ourselves into delusion.

I've seen ideas of religion both enhance and ruin lives. Despite my wavering, the most good in my life has come when I'm more in tune with my blurry notion of God and getting out of myself. So let's call it a draw.

Sunday after Mass, I ran into Brother Brendan outside the cloisters. "When people suffer bad things, they start praying," he told me. "And there's a reason for that. We intuitively know that there's something larger than ourselves, even if we don't know what it is." The research suggests he's right.

Scientists in Denmark tracked daily Google searches for prayer in ninety-five countries during the onset of the pandemic. The term hit the highest level ever recorded, growing 50 percent. The researchers say half the world prayed to end the coronavirus. They wrote, "The rise is due to an intensified demand for religion: We pray to cope with adversity." Even atheists, the research shows, are likely to lean on prayer when under stress.

And this isn't anything new. For example, after 9/11, 90 percent of Americans turned to religion. Even if we don't pray regularly, we do in the foxhole.

"People are natural searchers. We want a simple, true, universal outlook. We search and search for happiness, but often have a bunker mentality. We think it's in the next vacation or purchase. Me, me, me, hoard, eat, drink, acquire," said Brother Brendan. But, particularly when something bad happens, "we realize we should ask for help. So we pray. And it helps. Prayer helps. It's easy, it costs nothing, and it takes thirty seconds. Prayer is lifting your heart and mind to something larger than yourself. There's mental prayer, vocal prayer, meditation."

He begins telling me about the Catholic practice of praying the Rosary. "It's prayer where you're taking scenes from the Lord's life that have a supernatural, mysterious aspect and considering them. For example, like the night in the garden, where Jesus knew he would be crucified the next day and prayed for all of us."

The practice isn't entirely dissimilar to the Zen practice of koans. Koans are puzzling, paradoxical statements, anecdotes, questions, or verbal exchanges practitioners sit and ponder to aid meditation and gain spiritual awakening. For example: "When both hands are clapped, a sound is produced; what is the sound of one hand clapping?"

Koan practitioners admit that koans don't make any damn sense—

until they do. And when they do, they can lead to a mental shift that frees us.

"Prayer can only be done by humans," said Brother Brendan. "We can get grace from prayer. It can benefit our souls directly. It can change our souls for good, so all that we do immediately afterward will hopefully be better for others and, in turn, for us. It can help us be less self-obsessed and focused on the bunker mentality."

Meditation is like the Kim Kardashian of spirituality today, getting all the attention. But research shows both meditation and traditional prayer can deliver positive shifts in perspective, reduce stress, and help us better control our emotions. For example, researchers in Poland discovered that Buddhist meditators and Catholic prayers experienced similar positive changes in their brains' electrical activity. Both methods pushed them into the bands of brain waves associated with less stress and more contentment.

The research suggests that we should try what we believe works best for us. For example, one study found that stressed-out Catholics saw greater reductions in their heart rate and other stress measures when they prayed versus meditated. Meanwhile, a person drawn to meditation would probably do better with meditation.

After analyzing the research, I couldn't help but feel that it isn't *how* you're praying. It's *that* you're praying.

So whether it's a prayer to Jesus or Allah or meditating on koans or your breath or, as Mary Oliver explained her own practice, "I don't know exactly what a prayer is. I do know how to pay attention, how to fall down into the grass, how to kneel down in the grass, how to be idle and blessed, how to stroll through the fields." Just do something that helps you get out of yourself.

Once I'd put in a collective twenty sessions in the chapel, I discovered that I fall more into the Oliver camp. My favorite part of each day occurred at 6:45 a.m. As the brothers finished Psalms and lapsed into silent prayer,

I'd slip out of the chapel. I'd sit on the cloister stairs to watch the sun rising over the wilderness. I'd notice the birds being birds, the bugs being bugs, and the sun bending and illuminating a new day. I don't think that these brothers are wrong about their take on silence. Silence is when we hear.

Day five, 2:00 p.m. I had big plans to keep shoveling that pile of dirt and rocks. I was standing in the parking lot waiting for Brother Cajetan. My leather gloves were stuffed into the back pocket of my canvas work pants. Then Father Matthew approached me.

Father Matthew leads all the sessions in the chapel. He's the head priest at Our Lady of Guadalupe. And, I've concluded, the intellectual horsepower around here. One monk told me Father Matthew speaks six languages. "Let's go for a hike," he said.

As we crossed the parking lot in front of the chapel, Brother Cajetan skidded to a stop in the Chevy pickup. He was hanging out the driver-side window.

"I hear you're not working with me today!" he yelled.

"I'm hiking with Father Matthew," I said.

"Okay, have fun! That pile of dirt awaits us tomorrow!"

Brother Cajetan hammered the gas, kicking up dust and pebbles.

Father Matthew led us down a dirt road, past the monastery's fields, greenhouses, and chicken coops.

We then dipped onto a grassy trail, passing a thick pine stamped with a small and round triangular metal plaque. Big all-caps lettering arched around the top two edges of the plaque: "CONTINENTAL DIVIDE TRAIL."

Exactly 794.5 miles of the Continental Divide Trail's 3,028 miles flow north–south across New Mexico. We'd be tackling a few of the trail's miles running through monastery land.

The trail wound and dipped, losing elevation as it cut through the high desert. It was fatigued by Kelly green overgrown grass. "We've had a lot of rain this year," said Father Matthew. "The trail is usually much easier to see."

I asked him the same question I'd asked all the monks: "How'd you end up here?"

"I was at a university, and I was fascinated by the idea of money," he told me. "I took engineering classes because I thought that would get me the most money at the time. And I ended up getting two summer jobs that were very well paid, but sort of unexpected. Because of these two summer jobs, I had a guarantee for a super-high-paying job once I graduated."

This was in the 1980s in Midland, Texas, working in oil. "That time was a boom for the oil business," said Father Matthew. "It was such a boom that they had a Rolls-Royce dealer in Midland. Midland was a town of eighty thousand people. A map dot in West Texas's Permian Basin.

The Grateful Dead lyricist Robert Hunter in 1974 wrote the lines "Once in a while, you get shown the light / In the strangest of places if you look at it right." For Father Matthew, the light was glinting off the chrome hood ornament of a Rolls-Royce cruising through Midland's dusty roughneck oil fields.

"It seemed so blatantly artificial. Nobody I worked with actually wanted to be in Midland. But the only reason they were was for money," he said. The idea, Father Matthew explained, was that you would, until you hit sixty-three, do a job you didn't want to do, in a place you didn't want to be, so you could drive a car you paid multiple times the average American income for. And you would then show off that car to other people who were working a job they didn't want to work, in a place they didn't want to be, so they too could drive fancy cars.

"I realized that money was not going to satisfy my soul," he said.

Father Matthew looks like James Carville, but with the monk tonsure haircut. He was wearing a tan full-brimmed sun hat made of polyester wicking material. The type aging outdoorsy people buy at REI. His tan work robe hung to the tongues of his Timberland hiking boots.

He then began echoing Thomas Aquinas. "This search for possessions or titles or money is all a search for happiness. We convince ourselves that this next thing or accomplishment or meal or drink or promotion or raise is going to make us happy," he said. "And there's nothing wrong with any of these things. For example, if I have a drink to relax and to enjoy the company of some friends, that's no problem at all. If I make a purchase that will help me accomplish a greater goal, that's great. But these things can deliver a false and fleeting happiness. And we can chase that. Many people use these things and experiences to distract themselves and escape reality and numb something, and that's destructive."

Father Matthew is sixty, but he moves like a man half his age. No lag on the hills or hesitation underfoot. Bishop, the Benedictine happiness researcher, told me that most of the monks he's studied tend to stay at a healthy weight their entire life and remain mobile for most of their years. That's thanks to the *Rule*, which, remember, checks overeating and asks monks to move and work daily. "If you can keep moving and navigate your environment, that leads to more life satisfaction. It's a reason monks and nuns seem to have a higher quality of life and be happier as they age," Bishop told me.

After twenty minutes, Father Matthew paused at a small outcropping of chunky gray granite cooking in the sun. Around it a few prickly pear cacti emerge from the tan dirt ground. Beyond them a few piñon pines and burrobrush. Botanists at Western New Mexico University say there are more than two thousand species of plants in the Gila National Forest.

"That's where the rattlesnakes hang out," said Father Matthew. "They warm themselves on the rocks. They've been really active this year, so keep an eye out."

I heard a creek in the distance as we lost elevation. "We can convince ourselves that worldly things will make us happy," said Father Matthew. "When we get them, we feel good, but that happiness just doesn't last. Then we search for the next thing. We live in a big propaganda machine convincing us that the world is all about us. I can't tell you how many people I've seen who seem to have it all—a high-paying job, a nice home and cars, a spouse and family who love them—go through a midlife crisis and come here." Sometimes the monastery holds answers. Sometimes not. But the important part is that they've started asking the questions and searching for what strange place might show them the light.

"On the other hand, I've also met plenty of very wealthy people who are perfectly happy because they have something larger in their life," he said. Their wealth was just an afterthought. They happened to have a job that paid them relatively more than other jobs.

When we reach the creek, Father Matthew traversed black bedrock. It sloped and ran downward on both sides, funneling the river as a small canyon. The rock banks were smooth from years of water running atop them, carving deeper. He jumped across the river.

The stream surged past us, entering a wide pool that abutted a bank of black bedrock. We sat on that bank to continue talking.

I told Father Matthew my thoughts about how all we really know about happiness is that constantly ceding to the next craving—what the monks here might call "worldliness"—seems to breed unhappiness over the long haul. Once we understand that, happiness doesn't require as much as we might think.

"You see it here, at this monastery," said Father Matthew. "Look at all these young men here and the women at the convent. In spite of the austerity, in spite of the workload, in spite of all these things that might seem like a hardship and total waste of time, these people are all happy."

I thought of how taking on and overcoming challenges has been es-pecially rewarding to me. Especially intellectual challenges or big physical challenges in nature, like a backcountry hunt or long trail run. "It seems like doing hard things is rewarding to us," I said.

"Yes, that can be reward," he said. "On a natural level. But remember that can last only so long. You can't do that for sixty years and have nothing else but that. Your body will change. Challenge is part of it. An integral part of it. But challenge can't always be physical. We have bodies and souls. You also need to take care of and challenge your body and mind, but also soul."

A pinecone floated downriver. We watched it bob and bend with the flow and crash into a small eddy. Then he continued. "You're going to die—you need to face that," he said. "As healthy as you are right now, one day you won't be. Your body will fade. Then what? You're left with a soul. So you also need to focus on that. You need to find a deeper meaning. That's the thing. People focus too much on happiness. No one will ever be perfectly happy all the time because happiness is a moving target. It's better to focus on things we know are good and seek them. Then happiness becomes a by-product. Happiness comes by putting everything else in order and subordinating it to the ultimate goal. For us that ultimate goal is seeking God."

He's right that happiness is a moving target. Even the science of happiness is shaky, and we don't fully understand exactly what makes everyone happy. Yet we still chase what we hear will make us happy, then make decisions off questionable data.

Decades of research on happiness are inconclusive on exactly what ratio of our happiness comes from our choices. For example, one group of scientists reported that 50 percent of happiness comes from our genes, 10 percent from our life circumstances, and 40 percent from our choices. Which is to say, our actions determine 40 percent of our happiness. But other work shows that our choices account for only 15 percent of our happiness.

But who cares what the number is? The point is that science agrees that we *can* change some degree of our outlook. And it has to do with, as the Russian psychologist and sage happiness researcher Dmitry Leontiev wrote, "the way we organize our lives, relate to our fellows, the goals we pursue, with what is in our hands and is an object of our choice."

So then the question is, what are the right choices? What can we do to alter that 40 percent or 15 percent or whatever the percentage is?

Different studies and popular courses, books, and podcasts based on those studies push a mix of mindfulness, gratitude, and sociality. The monks of course practice prayer, an arduous form of mindfulness that differs from trendy mindful meditation.

They have a similarly counterintuitive approach to gratitude, which for them isn't about counting blessings, but sacrifice and scarcity. One monk told me that austerity and going without help them focus on what matters. And then, when they do receive a larger meal or other rare item that we take for granted every day, it becomes a blessing that they feel deeply grateful for. Occasional deprivation makes the ordinary feel extraordinary.

Modern research and thousands of years of wisdom indeed suggest that it's hard for humans to see and appreciate our blessings if they're always at hand. Purposefully going without can help us realize how great it is to have—to appreciate the wonders of our world of abundance. It's an idea embedded in ancient mythology and most religions. Consider Lent, Ramadan, or Yom Kippur.

We could occasionally spend time in the wilderness, totally removed from modern comforts. Or we could pick a comfort or two and go without it for a while. It acts as a hard reset. Then, when we get those things back, we can truly experience how wonderful they are. Gratitude often comes from scarcity, an idea backed by modern neuroscience.

We can also get this from giving, said Bishop. "Giving something you have that others don't is important for gratitude," he said. "I'm a firm believer that the people who live long, good lives consistently devote themselves to service."

But perhaps the biggest gap between modern happiness research and advice and the monks' way of living lies in sociality. Or, rather, their lack of it.

Yale University happiness researchers report that being around others is "a necessary condition for very high happiness." Like, no social? No happy.

The scientists point to research that finds social people are happy. But the monks at Our Lady of Guadalupe are something of an enigma. They're with each other, yes, but I don't know if I'd call them entirely social. They're muted most of their lives and essentially ignore each other for most of the day. The word "monk," in fact, comes from the Greek word *monos*, which means "solitary" or "alone."

Despite their lack of conventional sociality, these monks are happy. Bishop told me that social networks are important, but probably not the be-all and end-all as we're told. Especially as we age. "Monks have what's called anticipatory support," he said. "The monks may not talk to many people or interact with many people, but they know that if they need someone for help or just an uplift, they have a person. I think that is important for happiness."

And we can even build this anticipatory support internally. "We've found that the people who live past a hundred and are exceedingly satisfied with their lives tend to be religious," he said. "When all is gone, when all is lost, and not everything works for these people, they still have hope. They still have something they can seek solace in. It's like a survival mechanism for longevity. God is something they can rely on, talk to, trust, and feel like is there for them."

Of course it's beneficial to build good human relationships. But forced sociality may be counterproductive, and putting such a bright light on being social has likely led us to miss the other side of the happiness coin. "There's a difference between loneliness and solitude, and they're often conflated," said Bishop. "Solitude is purposeful and intentional."

The famed English psychologist Anthony Storr wrote of solitude, "Some of the most profound and healing psychological experiences which individuals encounter take place internally, and are only distantly related, if at all, to interaction with other human beings." His work showed that modern psychology idealizes human relationships so much that it can mislead people.

Forcing it can even create problems. Our social worlds and relationships often make us most miserable, as the work of the UC Berkeley researcher Cameron Anderson showed. Chasing human relationships for the sake of our own happiness can become its own hollow and pernicious scarcity loop.

Research and common sense suggest it's better to have one great friend you truly care about and can count on than a million mediocre ones. For some of the monks here, that "one great friend" does indeed seem to be God. There's a monk on the grounds of Our Lady of Guadalupe who lives as a hermit. He has a small cabin in the forest. He's maybe fifty-five years old and I've seen him only twice. He says nothing as he grabs a meal. Then he cheerfully skips back into the wilderness.

Others like these monks have existed throughout time. The people we see as the epitome of happiness often reached their state by shunning society. People retreat into solitude when they want to find enlightenment and total bliss. The Buddha did it. Jesus did it. Mary Magdalene, who is considered the apostle to the apostles, traveled to France after Jesus's resurrection and lived out the rest of her life alone in a cave. Saint Anthony, one of the earliest and greatest Desert Fathers, lived in solitude in the desert for more than thirty years. The Dalai Lama said, "To seek solitude like a wild animal. That is my only ambition." He argues that we need solitude to truly understand and change ourselves.

Uninfluenced time alone allows us to strip away outside noise and ask the deeper questions. It might even lead us to think differently. Better.

The eighteenth-century English intellectual Edward Gibbon said solitude "is the school of genius." He's not wrong. Isaac Newton, quarantined and alone from a plague outbreak in 1665, had his most productive years, revolutionizing our understanding of math and gravity. Gregor Mendel, a friar, discovered the beginnings of genetics while studying plants in solitude. Charles Darwin got to the bottom of his ideas on evolution after exiling himself at home alone after his five-year journey on the HMS *Beagle*. Nikola Tesla, arguably the greatest inventor of all time, said, "The mind is sharper and keener in seclusion and uninterrupted solitude.... Originality thrives in seclusion free of outside influences."

Tesla was onto something with his thoughts on originality and solitude. It also energizes creatives. Georgia O'Keeffe, Frida Kahlo, Emily Dickinson, Marcel Proust, Beethoven, Steve Jobs, and more have leveraged the creative power of solitude.

New research is proving them all right. A team of scientists at SUNY Buffalo recently found a major flaw in the idea that we *need* others to be happy. They analyzed the existing research and found that much of the work focuses on people who are alone because of fear or anxiety. In other words, these people didn't want to be alone, but were because of some psychological issue.

Is it all that surprising that asking a bunch of high-strung, terrified people who have no friends if they're unhappy would lead to some data that suggested having fewer friends can make one less happy? Nope.

With that in mind, the scientists studied groups they called "unsocial." These people don't have social anxiety or other fears and just prefer being alone. Give them a choice between going to a party or spending the evening home alone with a good book, and they'll take the party of one, thank you very much.

The scientists found that these people scored highest in creativity and seemed to be just as happy as their social counterparts.

I get this. Many of the moments I consider my "happiest" have been solo endeavors. Early morning hikes in beautiful places. Writing and thinking and weaponizing creative energy to make something.

Even though I choose these moments of aloneness, they can be trying at first. I've experienced lows of loneliness. But in suffering through that discomfort and attempting to uncover its deeper revelations, I've experienced insights. I've emerged from this inner abyss better for it—with greater self-knowledge and self-reliance. I've realized that I don't need another to feel okay. This process of becoming okay can be like the desert described in Jeremiah 2:6: "a landscape of steppe and ravine, a land of drought and danger, a land through which no one passes and where no human being lives."

But freedom lies in the austerity. When we return from this abyss, we're a better version of ourselves. More social, helpful, grateful, and empathetic. More appreciative of our good relationships.

When pressed, the happiness researchers recognize that millions of people throughout history have found immense meaning and satisfaction in the deep work of solitude. But they warn that solitude is not something for "average people." I disagree. The monks disagree. Thousands of years of myth and cultural tradition disagree. Who, exactly, is average?

By not exploring that abyss, we may prevent ourselves from rich levels of happiness, meaning, connection, and insight. It can help us become extraordinary. As an unnamed monk wrote of solitude in the nineteenth century, "You risk so much by hesitating to fling yourself into the abyss."

Which is all to say that happiness is a moving target. We understand what makes some people happy, but not everyone. When we chase the newest thing we hear will make us happy, we make decisions off questionable and ever-changing data. It's as if we were always overhauling our life to abide by the finding of the latest happiness study.

Father Matthew looked at the horizon and said, "We just put everything in our lives in order and subordinate it to our ultimate goal, which is seeking God. If you're only focused on you and your happiness, you're going to crush others, first of all, and you're going to destroy your own happiness. The way we end up happy is forgetting about ourselves and loving God."

Of course, God isn't the answer for everyone. Not even close. But perhaps it all comes back to realizing that we aren't the center of the universe. That there are things greater than us. We can't necessarily quantify these things, and they aren't found in fleeting pleasures or fame or followers or money or stuff or apps. Our well-being seems to be determined not by any one end point but by a rolling average of our actions. And also a willingness to explore our innermost selves rather than scramble after the next thing we think will make us happy. Spirituality counters what scarcity brain pushes us into and asks us to do the deep work.

It's getting out of a happiness scarcity loop by evaporating the opportunity. The opportunity we're seeking is no longer "to be happy." Seeking happiness in and of itself, as Father Matthew said, can backfire. The opportunity instead becomes a range of outcomes that are helpful to ourselves and others. And as all the unpredictable rewards from our actions accumulate, we find ourselves happy.

Father Matthew checked his watch. Vespers began in forty minutes, so we rose and hiked the winding trail back to the monastery. When we arrived at the chapel, he stopped and turned to me.

"If you don't mind me saying, I think you're seeking," he said. "All your traveling and learning, you're not going to find what you're seeking anywhere on earth. You need to break free of time and space and find something larger. That search has been going on since the beginning of man."

He continued: "Ask, what should I do with my life? Where am I supposed to go? What am I supposed to be doing? Who am I? What does all this beautiful order mean? Those are questions you have to ask yourself seriously and answer. You're the only one who can do it. Ultimately, it's you and something larger. That's the drama of all human life. And it's worth being a part of."

So perhaps happiness is the dramatic effort of a long and hard walk with seemingly no destination. The terrain is rough and the weather isn't always perfect. It's a stroll into an abyss. But at some point along the way, by trying to make each step a bit less about our immediate desires, we realize we're happy, even though the journey hasn't ended.

What We Do Now

On my final day at the monastery, I entered the chapel for Matins. The monks all filed in at 3:25 a.m., as they do every day. As they will for the rest of their lives.

"In Deo salutáre meum, et glória mea: et spes mea in Deo est."

I listened to their chants reverberate off the stone walls in the dim light. A cold breeze was pouring in from the cracked windows and door. After fifteen minutes, I rose and returned to my car, winding it down that long dirt road once again, heading back to Las Vegas.

I took all of the nine-hour drive in silence. It was a good time to start processing my last two years investigating scarcity brain and how it changed me.

I was on an empty two-lane road that intersected rolling arid plains. The sun was rising in my rearview mirror, rendering the sky behind me entirely gold. I thought back to a conversation I had early in the journey.

Before I left for Iraq to research addiction, I flew to San Diego to meet Mike Moreno. Moreno now works with technology start-ups. But he spent more than a decade in the Middle East for various government agencies, including years as a CIA operations officer. Much of that time was in Iraq at the height of combat.

He'd agreed to give me a daylong crash course in survival. He was the right teacher for my situation.

He picked me up at the airport and tossed my pack in the backseat of his SUV, which held a duffel bag filled with duct tape, handcuffs, rope, zip ties, hoods, and guns.

Moreno and I drove past the harbor and headed inland, deep into the rugged and hilly desert of the U.S.-Mexico border. Up and down dirt roads flanked by dried brush and engulfed in the unrelenting heat. We were on the U.S. side, ten miles outside Tecate, when we found a small canyon where few people go.

Soldiers are heavily armed, coated in body armor, and embedded with a group of other soldiers who are also heavily armed and coated in body armor. They have safety in weapons and numbers.

"But what you're doing in Iraq and what I did for the CIA, they are very similar," he said. "We both work alone and mostly defenseless in kinetic environments to acquire information from sources. I know you can get information from sources. But you need to learn how to survive if something goes wrong."

Moreno covered deep theory on the art of avoiding danger and kidnap. But also how I could possibly escape if kidnapped.

First, I learned soft skills like how to interact with locals. How to behave so possible captors might begin seeing me as a human being. "ISIS is not just going to let you go," he said. "But saying and doing the right things at the right moments might get them to do something like slightly loosening your bindings. And that could present an opening."

Then we covered the hard skills. How to escape if my hands were bound in front of me or behind me by handcuffs, zip ties, or duct tape. How to react if someone pointed a gun in my face or came up behind me and held it to my head. How to break out of the trunk of a car. How to effectively use a series of urban survival tools he'd told me to buy. How to increase the odds of surviving a gunshot.

Our conversations were occasionally interrupted by the loud chop of helicopter blades. U.S. Border Patrol regularly scans the area from above. At which point we'd realize that what we were doing probably looked like at least a Class D felony. Moreno would conceal the fake gun he was holding. I'd hide my bound wrists. We'd both laugh.

Eventually, we were sunburned and sitting on the tailgate of his SUV. "All of these skills we've just covered," Moreno said, "you're never going to have to use them—until you do. And then your life will depend on them."

He continued: "But what I really want you to walk away with is knowing that if something goes wrong, your survival will come down to a will to live. Some people just give up. But the people who escape and survive have the will to live. They never quit. They never give up. They cling to life, push through pain, and summon effort and never stop making the choice to do the hard things that are pushing them in the direction of survival."

And that is, he said, ultimately a choice. Then he paused.

"I'm honestly jealous you're going to Iraq," he told me. We'd spent eight hours covering all the reasons my journey there wasn't exactly favorable for longevity. "Why?" I asked, somewhat incredulously.

"Well," he said, "there was a purity and focus of life and mission there. The extremes of the human condition and human experience put life into such sharp focus. It recalibrates you in a way you can never come back from. It reminds you what's really important in life."

I'd read similar accounts. For example, in *The Things They Carried*, Tim O'Brien wrote of his experience as a soldier in Vietnam, "At its core, perhaps, war is just another name for death, and yet any soldier will tell you, if he tells the truth, that proximity to death brings with it a corresponding proximity to life. After a firefight, there is always the immense pleasure of aliveness."

I later asked my friend Jaala Shaw about this. She's spent years teaching children in conflict zones like Afghanistan. "These austere environments are my favorite places on earth," she told me. "Not because of the conflict. But because of the meaning people find and I find in life despite the austere circumstances. Some people think I hunger for danger. That's not it. It has more to do with actual living. You learn to live more meaningfully when you can die at any moment."

Another friend, Yohance Boulden, a soldier who did five tours in Afghanistan and Iraq, echoed the same. "I don't think I'll get over my time there," he said. "In war, things we take for granted become exciting and you appreciate them. You get stoked for mail, hot food, hot showers. You feel most alive and focused when decisions feel like they have the ultimate consequence."

When I returned from Iraq, I called Moreno. "I get it," I told him. I explained how there was something counterintuitively life giving about the present tense of living that Iraq forced. My head was always on a swivel. Each action felt consequential.

And I was there for only one week. I understood the longer I stayed, the higher the risk. Not just physically, but also psychologically.

"I have a lot of guilt wrapped up in how I feel about my time there," Moreno told me. "The circumstances there definitely took a toll on my mental health, but also made me a better and more appreciative and useful person. It completely changed what I consider a problem."

There's a thin line between post-traumatic stress and post-traumatic growth. Some people who have experiences like Moreno, Shaw, and Boulden never really return.

"I just don't know how to square that circle," Moreno said of his complicated feelings. "I don't want my kids to go through what I did. But they need to have experiences in life so they have perspective and don't take things for granted. And, of course, it's not just my kids. It's everyone."

Driving home from the monastery, I thought a lot about the will to live. For nearly all of time our ancestors had to practice the will to live. Each day we'd struggle for more, powered by the scarcity loop to persist. And when we'd find what we were looking for, we'd experience deep rewards. As Zentall said, "That extra psychological value encourages future persistence and energizes us to keep looking. I think that gets translated in the modern world in which humans value things they had to work longer and harder to get."

But these scarcity loops of our past were different from those offered to us now. We craved pleasure and ease, but we rarely got them. Our scarcity loops were challenging and uncomfortable in the short term but rewarding in the long term. The will to live was forced.

The great miracle of living today is that our survival no longer requires constantly exercising the will to live. But there's also peril embedded in that promise.

Boulden told me, "What I struggle with most today is people complaining about trivial things or wasting their time. People wasting their time really bothers me. People label it as me just being an irritable vet. But I think it's a case of seeing each moment as precious and knowing that I'm watching lost time." This is a common refrain among veterans of war. And after my travels into extreme and austere environments, I see it in myself. We have one shot at life.

The scarcity loop, as it too often exists in our modern world, has flipped. Today it pushes us into short-term comforts like mind-altering substances, online diversions, and acquisition. These things are all fine and fun used consciously and in moderation. But we can too easily escape into our modern scarcity loops too often for reasons other than fun—at the expense of long-term rewards, growth, and meaning. And that's when the problems bubble to the surface.

These are physical, psychological, and even spiritual problems. Yet it's hard to exit the loop, because it feels good and offers a short-term escape from the problems it causes. And it's so natural to us. We're doing what our species has always done—except on a radically new playing field. These are the waters in which we swim. Zentall's caged pigeons only gamble because they need something that feels like exercising the will to live.

The people I met on my journey are a lot like Tiktaalik, that ancient fish that left the water for something else. Or like Zentall's pigeons who were freed into wild environments and stopped gambling. Or like Moreno, Shaw, Boulden, Zerra, Vande Hei, and the monks, who all put themselves in positions that forced the will to live.

Moreno said, "How can we pass on the lessons of these experiences without people having to actually go through them? I think part of it is taking small risks in everyday life. Seeking adventure and experiences that build perspective."

These risks and adventures can be both big and small. Problems from escaping into the scarcity loop are often a signal. They indicate that it's time to make uncomfortable choices that force the will to live.

The slot machine engineers in Las Vegas told me that a person stuck in a scarcity loop only stops for three reasons: the opportunity goes away, the rewards stop trickling in, or the repetition slows down.

We learn to fall into addiction when a substance or behavior gives us an easy opportunity to solve our problems and improve our lives for a fleeting moment. But we can, as Dr. Abdul-Razaq told me, work hard to solve our problems in a different way and create new and better opportunities for ourselves.

We can resist the pull of numbers that boil complex human experiences that are beautiful in their complexity down to a simple game. We can determine for ourselves what we want the rewards of a behavior to be. They don't have to be likes, follower counts, grades, salaries, rankings, and scores. We can invent our own game with different goals that enhance how we spend our time and interact with others. These goals will be harder to measure, but they'll be far more meaningful than anything we can quantify.

We can slow down the quick repetition embedded in our modern food system by eating Tsimane-like foods. The foods humans have been eating for thousands of years. And, if we must, remove the opportunity entirely by not keeping the foods that trigger us in the house.

We can reduce how frequently we buy by viewing our purchases through the lens of gear rather than stuff. It's taking a utilitarian mindset and applying it to our current and future possessions to accomplish more of what truly enhances our lives and gives us meaning. Having items "earn their weight" and creatively solving a problem with what we have are deeply rewarding and often lead to better outcomes. Once we experience those rewards, the unpredictable reward a future purchase may bring becomes less alluring.

We can resist our informavore brain and its craving for certainty and answers to every unknown. We can realize that the unpredictable rewards of information we read online are often a misdirection. They don't always improve our lives or lead to understanding and often stress us out. We can do what humans of the past did: exercise our drive to explore; break new ground on the map and in our minds. We can get out into the world and work to know the unknown. That's what leads to wisdom, understanding, and a storied life.

We can shift scarcity loops into abundance loops. That is, find hobbies that have the three parts of the scarcity loop but help us do things that are good for us. Like what Zerra does with her shed hunts or what Hanke did with *Pokémon Go*. Many activities in nature thrust us into the loop. For example, fishing, bird watching, rock collecting, and more all mimic the scarcity loop. No, the repeatability isn't as rapid as we'll find online in TikTok or slot machines—we won't be catching fish or seeing bald eagles every two seconds—but the harder work the activities require can make them more rewarding. Along the way, we're doing what's always been good for us.

Or this could be exploring a new place and seeing what unpredictable rewards it might hold, as simple as trying a new restaurant without googling it first. Or it could be shifting our search for happiness to something tangible that improves our happiness along the way, like helping others, doing the next right thing, or understanding ourselves better.

We can also use the scarcity loop to get more of what we want. We can use it in the workplace to keep our fellow workers engaged. For example, the famed behavioral psychologist Karen Pryor noted that employers who throw unexpected special lunches for their employees generally get better feedback and more loyalty from workers than those who put special lunches on the calendar. It's the same lunch—but unpredictable rewards are more exciting. Or if we want our child to do a good behavior, we can regularly reward the child early, but then slowly shift to unpredictable rewards. Remember that most species become bored of getting the same reward every time they do a behavior. It becomes work. Doling out random rewards is more likely to keep people engaged. So, next time your kid cleans his or her room, give them a reward—every time at first, and then only sometimes.

We should look for the loop whenever we find ourselves mired in a bad behavior. Take sticking with bad relationships. As Pryor explained, "If you get into a relationship with someone who is fascinating, charming, sexy, fun, and attentive, and then gradually the person becomes more disagreeable, even abusive, though still showing you the good side now and then, you will live for those increasingly rare moments when you are getting all that wonderful reinforcement: the fascinating, charming, sexy, and fun attentiveness." It leads us to stick around too long thanks to the power of unpredictable rewards. But once we realize what's happening, it becomes easier to walk away.

Improving our lives still requires enduring short-term discomfort for long-term achievement. It should still guide our actions today.

Our scarcity brain naturally resists this; doing the harder thing hasn't made sense until now. The monks taught me that wading into this murky abyss of asking the tougher questions about ourselves and intentions and letting the answers guide us can be uncomfortable, frustrating, and sometimes dark. But being willing to go there and, despite the layers of challenge, summon the will to live makes life worth living. It's the human story. After analyzing thousands of years of mythology across cultures, Joseph Campbell explained the story like this: The cave you fear to enter holds the treasure you seek.

When I considered the experiences that most changed me for the better—experiences that made me more appreciative, present, empathetic, and helpful—all of them were difficult. When I went through the hell of getting sober, my life improved across the board. I had to relearn life and living it and work to fix the external and internal wreckage of my past.

And it's been an ever-evolving process. When I sat by the side of the creek with Father Matthew and told him how it's been rewarding to challenge myself physically and intellectually, he helped me unpeel another layer of the onion of living. "Yes, that can be rewarding on a natural level. But remember that can last only so long . . . and it can ultimately be destructive," he said. "You also need to take care of and challenge your body and mind, but also your soul."

He helped me see that when I left the scarcity loop of addiction, I fell into a new sort of scarcity loop. One of perfectionism. I took up hard exercise, became obsessed with work, and more. I became terrified of failing or looking bad—revealing any remaining wreckage. In this scarcity loop, I felt the short-term satisfaction and clarity of measuring my progress in numbers like salary, accolades, and book sales.

My journey to understand scarcity brain led me to start trying to square that circle of living. I've begun unpacking why I kept chasing more in the first place—from alcohol to accolades. It's often agonizing work. The hammer is heavy, and the circle isn't all that pliable. But more is revealed as it bends. I'm slowly wading into that uncertain abyss and finding deeper things about myself I didn't know were there.

But that abyss has forced me to exercise the will to live in new ways. And I've started to love the weight of the hammer. Because maybe the point of hammering isn't to actually turn the circle into a square. Maybe it's to improve as a human as you hammer. To callous your hands, build endurance, and improve your craft of life and living it.

And maybe it's the same for you, too. When I was in Baghdad with Ehab and Nader, the two Iraqi intelligence officers, they asked me why I was speaking to them.

"I'm writing a book," I said.

Ehab smiled as he blew lemony tobacco smoke from his mouth. "You won't sell many books here," he said. "We don't read many books. We have problems. Books are for people who have relaxed lives." We all cracked up laughing.

He was, of course, joking. But I think of his joke often because it holds an element of truth.

As the Las Vegas Strip came into sight through my car's windshield, I remembered the words that unnamed monk wrote in the nineteenth century. "You risk so much by hesitating to fling yourself into the abyss."

ACKNOWLEDGMENTS

Thanks to my wife, Leah, for all your help and humor as I worked on this project. I appreciate you listening to my incessant streams of thought, reading the first drafts of this book, and helping me improve it. The work is better thanks to you.

Thanks to my two dogs, Stockton and Conway, for helping me remember rule 62: Don't take yourself so damn seriously.

Thanks to my mom, the badass, for all your support and for encouraging me to pursue writing.

Thanks to my editor Matthew Benjamin. We have now entered cranky old couple territory, and it's made our books better. Here's to getting even older and crankier on future projects.

Thanks to my literary agents, Jan Baumer and Steve Troha, for reading my words and having my back.

Thanks to all my friends who read early versions of this work and offered wisdom. Most important, Trevor Kashey, for his Dr. Evil-like, brilliant notes. You helped me get a handle on so many of the themes in this book and I'm forever thankful for your thoughts and friendship.

And Bill Stump, Bill Stieg, Ben Court, Ebenezer Samuel (who secretly enjoys CrossFit, I hear), Cynthia Shumway, Jason and Emily

McCarthy, Tom Thayer, Brady Holmer, and Tyler Daswick. The book is better thanks to all of you.

Thanks to all the people in the pages of this book who were willing to speak with me, host me, and answer my incessant questions. I'm grateful to Thomas Zentall, Daniel Sahl, Thi Nguyen, Sally Satel, Maia Szalavitz, Kevin Hall, Rachel Laudan, Stephan Guyenet, Mike Roussell, Michael Gurven, Mark Vande Hei, Stephanie Preston, John Hanke, Caroline Rose, Alex Bishop, and so many more. Special shoutouts go to my friends in Iraq and the Amazon, Mike Moreno, and Laura Zerra for keeping me alive in strange places. And, of course, the monks at Our Lady of Guadalupe for compelling me to peel the onion of living.

Finally, thanks to everyone who spends their time and attention reading this book. I hope you get something from it that enhances your life.

INDEX

ABOUT THE AUTHOR

MICHAEL EASTER is the author of *The Comfort Crisis* and a professor at UNLV. He writes and speaks on how humans can leverage modern science and evolutionary wisdom to perform better and live healthier and more fulfilling lives. His work has been implemented by professional sports teams, elite military units, Fortune 500 companies, leading universities, and more. He lives on the edge of the desert in Las Vegas with his wife and their two dogs. To get new ideas and tips to improve your life, subscribe to his free *2 Percent Newsletter* at eastermichael.com/newsletter.

Also by
MICHAEL EASTER